GENETICS
Journey from Basic to Biotech

GENETICS

Journey from Basic to Biotech

Dr. G.C. Satpathy

KALPAZ PUBLICATIONS
DELHI-110 052

GENETICS: Journey from Basic to Biotech
(Applied Science)

Rs. 825

ISBN: 81-7835-183-8

Published in 2004 in India by
Kalpaz Publications
C-30, Satyawati Nagar
Delhi-110 052
E-mail: kalpaz@hotmail.com

Lasertypesetting by: Varsha Graphics, Delhi
Printed at: Mehra Offset Press, Delhi

Contents

Message

I am happy to know that Dr. G.C. Satpathy, Chief Medical Officer, Paradep Phosphates Limited is publishing a book on 'Genetics'. The knowledge of Genetics is imperative in the field of Medicine, Veterinary and Agricultural Science.

I wish him all success in his endeavour.

K.K. Gupta
Managing Director

Foreword

Last century witnessed spectacular advances in Science and Technology and brought new frontiers in science and new technologies making communications much easier and faster, understanding of the biosystems, plant, animal and human behaviour and response to different stresses and understanding the human genome etc. These advances in Science and Technology have opened up new branches like Information Technology, Biotechnology, Material Science etc. which are believed to become the science of 21st century.

Genetics is a discipline of Biological Sciences whose foundation was laid by Charles Darwin, Greger Mendel in the 19th century and the science developed in the 20th century to such a position that man could create transgenic plants and animals for protein production and increased food production through Green Revolution and control of pests etc. Genetics forms the basis of modern Molecular Biology and Biotechnology. The medical and social implications of plant genetics and human genetics are enormous.

Dr. G.C. Satpathy, M.D. has written a book on Genetics covering a historical account, Mendelian principles and cell division, linkage and crossing over and gene interaction, sex determination and sex-linked inheritance, chromosomal aberrations, polyploidy and non-chromosomal inheritance in a very simple and lucid manner for college and university students. Dr. Satpathy has used insects, fowl, mice, human systems and plant systems to explain the complex mechanisms involved in inheritance in a simple and illustrative language. Although the book does not deal with modern aspects of genetics involving advances in Molecular Biology and Biotechnology but it provides enough useful materials for the students to understand the Science of Genetics and problems involved in it.

I enjoyed reading this book and I am very much hopeful that the book will be liked by college and university students, medical and veterinary scientists and agricultural scientists. Dr. Satpathy's efforts are praiseworthy and I have great pleasure in writing this foreword for the book.

Prof. M.C. Dash
Vice Chancellor
Sambalpur University,
Jyoti Vihar, Sambalpur,
768 019, Orissa

Preface

Five decades ago, James Watson and Francis Crick proved that our destiny is not written on the forehead but is beautifully coded in a string of DNA molecule the blue print of life—housed in the nucleus of all our body cells. This year 2003, we are celebrating the Golden Jubilee of discovery of DNA. But prior to this historic event, scientists all over the world have tirelessly researched to understand the basic principles of inheritance—called Genetics. As the time passes, when new frontiers of scientific inventions were brought to light, there has been significant revisions on the basic concepts on the earlier knowhow on this interesting and ever expanding science. All biological, biochemical and pathological processes are explained in the light of genetics in medical, veterinary and agricultural fields, coming in big way as an useful tool to answer the age old questions.

The present book is aimed at highlighting some of the revised concepts in the basic genetic principles keeping adequate distance from the silver lining of newest inventions on the field.

I record my sincere thanks to large number of benefactors who have extended their unconditional co-operation on the compilation job of this book. I am greatly indebted to my teacher Prof. M.C. Das, Vice Chancellor of Sambalpur University, an eminent biologist and environmentalist, who has scanned the manuscript and written the 'Foreword', I am also grateful to Mr. K.K. Gupta, Managing Director of one of the leading fertilizer manufacturing concern in the country for his complementary message.

My wife and children also play a vital role as boosters for my "clinico technical word mixing."

—Dr. G.C. Satpathy

Introduction

Genetics is the branch of science that deals with the transmission of parental characters to their offsprings including usual vexations. It has achieved a prominence among the biological sciences in recent years for laymen and scientists alike. Laymen are now more aware of the medical and sociological implications of human genetics; physicists and chemists continue to find fascinating problems in fundamental genetics that they can approach and in some cases solve with their different kinds of expertise; psychologists are finding that the study of the inheritance of behavioural patterns in animals is providing new insight into the understanding of behaviour; medical scientists have become aware that a deeper understanding of many diseases will only be attained with further knowledge of their genetic bases; and ecologists, confronted with the enormous complexity of the environment-organism relationship, now appreciate that an important component of this interaction is a genetic one. Modern genetics is a complex of inter-related subsciences that pervades all of biology and provides a major part of the conceptual framework of modern biology. This statement is not an exaggeration; one need only consider the questions that geneticists ask to recognise its validity: What is passed from parents to offspring that determines the characteristics of the offspring? What is the mechanism of this transmission? Why are the offspring of sexually reproducing parents somewhat different from the parents? What factors guide and direct the development of a fertilized egg into a highly differentiated multicellular adult? What is the source of the variability in a population of a given species of organism? How is one species different from another species? What constitutes a species? How do organisms evolve? These are among the major problems of biology, and, although not all of them have beer answered, it will be biologists using the tools of genetics who wi provide many of the answers.

Thus, there are geneticists who are biochemically oriented and work at the molecular level, those who work at the cellular level with chromosomes and other cellular organelles, and those who work with populations and are mathematicians or ecologists. Some even use the tools of paleontology to try to unravel what has happened in the past. Others are crystallographers who are unraveling the architecture of macro-molecules that determine the structure of the organism and contain the information that is passed from one generation to the next. This heterogeneous assortment of methodologies and approaches constitutes genetics and makes it hard to delineate the scope of the field. However, there exists a "core' of knowledge drawn from this heterogeneity that forms the principles of genetics. In this book we attempt to make clear what this core is, principally for the eukaryotic organisms—those organisms with true nuclei in their cells as distinguished from the prokaryotes such as the bacteria that do not have nuclei.

The Beginnings of Genetics

The nineteenth century was an exciting period in the development of ideas in the western world. Many concepts that form the framework of the physics, chemistry, geology and biology of the present arose—and to a great extent were developed—during that century. Three events in the 1800s in the area of biology were to be important in dictating the course of its development up to the present. First, Charles Darwin published his *Origin of Species;* second, the cellular basis of plant and animal life began to be investigated and its significance appreciated; and third, Gregor, Mendel published the results of his experiments with pea plants. Modern genetics owes its origins to all three of these events.

Classical genetics. Mendel's work, although published in 1866, was virtually ignored until 1900 when it was rediscovered. Let us start with what he did in the 1860s. Mendel interpreted the results obtained from crosses with pea plants as support for the idea that heredity has a particulate basis. Implicit in this hypothesis is the existence of physical elements in germ cells that pass unaltered from generation to generation through the intermediate adult forms. Mendel had no clear conception of what these elements were. At that time cytological knowledge was in a primitive state. Cells were recognised, but chromosomes were not, and the significance of nuclei was not realised until the 1880s. For the present it is only necessary to know that the significance of Mendel's discovery was not realised until 1900. In that year, Hugo de Vries and Carl Correns published papers in which they reported repeating the types of experimental crosses made by Mendel. They came to the same conclusions he did. This time the full significance of the results and the conclusions to be drawn from them were recognised by most cell biologists and by many plant and animal breeders. Interest in genetics became intense, and the science developed, explosively.

One of the reasons for the rapid advancement of Mendelism was that by 1900 cell biology had reached a critical stage in its development. Chromosomes were by then recognised as the bearers of the hereditary material, the significance of mitosis was appreciated, and the nature of meiosis was beginning to be understood. By 1903 it was realised that meiosis with its separation of homologous chromosomes, and the formation of gametes followed by fertilisation, provided an explanation for the results of Mendel's experiments. This union of cell biology with Mendelism between 1900 and 1905 might be called the beginnings of 'classical' genetics—the genetics of the first half of the twentieth century. This period saw the development of the *chromosome theory of heredity*, and the working out of the mechanics of inheritance in several eukaryotic organisms.

During the latter half of this 50-year span, genetics began to provide the logical basis and experimental evidence for Darwin's theory of evolution by natural selection. Darwin had no satisfactory explanation of heredity, and he did not understand the sources of natural variation. Hence, the underpinnings of his theory of natural selection were absent. Genetics provided these underpinnings, and the result of this fusion of Darwinism and genetics was *population genetics* and a further development of evolutionary theory.

Molecular genetics. In 1869 Friedrich Miescher, a young Swiss physician, announced the isolation of nucleic acids from the nuclei of pus cells. Like Mendel's discovery, the significance of Miescher's work was also not immediately appreciated. At the turn of the century, the American cell biologist, E.B. Wilson, did observe that nuclein was probably the essential component of the chromosomes, and hence the hereditary material. However, this prescient observation aroused little interest, and it was not until 1944, 100 years after Miescher's birth in 1844, that experiments were reported which strongly indicated that deoxyribonucleic acid, the main component of Miescher's nuclein, was indeed the hereditary material of certain bacteria. The recognition that something physical had to be passed from parent to offspring was firmly accepted in the last part of the last century, but what it was no one knew. The demonstration that this physical factor was a nucleic acid, the structure of which had already been partially worked out by the 1940s, ushered in a new era not only in genetics but in biology as a whole. This was the beginning of molecular genetics that is part of the broader field of molecular biology. The fusion of classical genetics with molecular genetics and the genetics of the lower organisms (prokaryotes) has evolved into modern genetics.

The Scope of Genetics

Genetics is *the study of genes through their variation.* A ***gene*** is the basic functional unit of heredity, which is composed of a section of a long, threadlike biological molecule called ***deoxyribose nucleic*** acid, best known by its abbreviation ***DNA.*** In the nuclei of higher organisms are found the ***chromosomes,*** each a very long, single, continuous DNA molecule containing thousands of unique functional units, the genes, as part of its length. Each of the trillions of cells in a human being has 46 chromosomes, in two equivalent sets of 23. Each of the 23 in a set is unique and is matched only by its equivalent partner in the other set. An average human chromosome contains about a 50-millimeter length of DNA, and, therefore, one total chromosome set of 23 is the equivalent of about one meter of DNA.

The study of the modes of gene transmission from generation to generation is broadly called ***transmission genetics,*** the study of gene structure and function is called ***molecular genetics,*** and the study of gene behaviour in populations is called ***population genetics.*** These three form the major subdivisions of the field of genetics, although, as with all categories invented by humans, the subdivision is to a certain extent arbitrary and there is considerable overlap. It is the knowledge of how genes act and how they are transmitted down through the generations that has unified biology; previously, specific sets of biological phenomena had each been relegated to separate disciplines.

An understanding of how genes act is now an essential prerequisite for such biological fields of study as development, cytology, physiology, and morphology. An understanding of gene transmission is a fundamental aspect of areas such as ecology, evolution, and taxonomy. Further unification has resulted from the discovery that the basic chemistry of gene structure and function is very similar across the entire spectrum of life on the earth. These points may seem trite to those who have grown up in the light of current knowledge, but it is important to realise that our modern view of biology and its interrelated parts is a relatively recent phenomenon. Not so long ago, biology was fragmented into many camps that rarely communicated with each other. Today, however, every biologist must be a bit of a geneticist, because the findings and techniques of genetics are being applied and used in all fields. Genetics, in fact, provides the modern paradigm for all of biology.

What have been some of the success stories of genetics within basic biology? For just about any feature of biological structure or function—such as size, shape, number of parts, colour, pattern, behavioural pattern, or biochemical function—that has been looked at in experimental organisms, genes have been found to be involved. Determining the location of these genes on their respective chromosomes is relatively easy. It has been found that the major way in which genes exert their effect is through controlling the myriad chemical reactions that go on inside cells, and thousands of specific genes have each been associated with a specific chemical reaction.

The control of gene action has been intensely studied in some organisms, and the mechanism in these cases is well understood. Many different genes have been isolated in the test tube, and their particular chemical structures determined. Such studies have provided important clues about how genes perform their functions. A gene can be removed from one organism and introduced into another, either for the convenience of propagating large amounts of the gene for later study or to examine its effects in another biological system. Genes can be modified at will to study the effects of these changes on biological processes; such changes can involve a large part of the genes, or very localised regions of interest to the experimenter.

Most genes have been found to reside rather stably at specific chromosomal locations, but other pieces of DNA have been found to be capable of sudden relocation to new areas. Last but not least, most of these findings have tremendous relevance for evolutionary processes, which of course are concerned with changes in the structure and function of the gene set. Clearly, the advances in genetics have been truly astounding, particularly over the last three decades. Many recent accomplishments—such as the isolation and characterisation of individual genes, which researchers in the 1950s believed could never happen in their lifetimes—have already come to be regarded as routine procedures in current work in genetics.

The Flow of Genetic Information

The flow of genetic information in all organisms appears to be basically the same: replicating a nucleic acid and passing the replicas on to the next generation. In viruses this nucleic acid may be either RNA or DNA. In prokaryotes and eukaryotes it appears always to be DNA. An important part of the study of genetics is the mechanisms of this transfer and their consequences. The mechanisms differ from one organism to the next. The viruses and prokaryotes may pass on all of their nucleic acid or only part of it; indeed, only fragments may be transmitted. The eukaryotes pass on whole chromosome sets through the processes of mitosis, meiosis, and fertilisation. The transmission of the genetic material and its reassortment to form new combinations is of the utmost importance to all organisms. A vital part of this study of transmission is its molecular aspects. It is important to understand the chemical nature of the transmitted material, which then leads to understanding the mechanism of its replication and its functioning as an information system.

The Genetic Material

When it was recognised that the nuclei of cells must contain something inherited, presumably the chromosomes, which functioned to determine the characteristics of organisms, questions were raised about its structure and function. The answer to these questions did not begin to be realised until the development of molecular biology in its modern form in the 1950s and 1960s, an important part of the study of genetics as a whole. It encompasses areas relating not only to the primary function of nucleic acid as the genetic material but to the mechanism of regulation of its functioning (since it does not all function at once at any given time). This then relates directly to the problems of differentiation and development. The tremendous variety seen in the organic world has its sources in the differences in the genetic structure of the different organisms. The origins of these differences and their various affects on the characteristics of organisms is another important part of the study of genetics.

The Genetic Structure of Populations

Genetics is concerned with the genetic structure of populations, as well as that of the individuals who make up the populations. The study of population genetics is a vital part of genetics, because the variations that arise and are preserved in the DNA are the material on which natural selection operates to bring about changes in populations over periods of time. Knowledge of the genetic structure of a particular population can lead to predictions about the genetic structure of future generations. This ability to make predictions about future generations is becoming more and more important in the area of medicine.

The Unity of Genetics

In the first part of this section we have attempted to give the reader a feel for the broad scope of genetics as it relates to the more general but fundamental aspects of biology. Actually, if one were to make a list of the different fields of genetic research and application, only some of the terms used to describe the broad scope would be used. Terms such as bacterial genetics, wheat genetics, phage genetics, and the like, would be applied to indicate orientation toward a particular organism. But in each of these the same concepts and approaches as have been described above would be used by workers with specific organisms. Fundamentally they all ask the same kinds of questions; this is what makes genetics a coherent science despite its great complexity. Other new fields of genetics are coming into existence—which instead of being organism oriented are specific problem oriented.

Genetic Analysis

With the wealth of genetic knowledge we now share, we can make generalisations about DNA, genes, phenotypes, and genotypes as though these concepts were self-evident. But obviously this was not always the case; the current wisdom was acquired only after extensive genetic research over the years. Mendel, for example, almost certainly was completely without a conceptual basis for his research at the beginning of his work, but he was able to piece together his genetic principles from the results of his many experiments. The process of identification of the specific hereditary components of a biological system is called ***genetic dissection.*** A geneticist is a type of biologist interested in some aspect of the structure or function of organisms. In the same way that an anatomist probes biological structure and function with a scalpel, the geneticist probes biological structure and function armed with genetic variants, usually abnormal. If the geneticist is interested in a biological process X, he or she embarks upon a search for genetic variants affecting X. Each variant identifies a separate component of X. In much the same way that a novice auto mechanic can learn a lot about how an internal combustion engine works by pulling out a spark plug lead, for example, the geneticist "tinkers" with a living system.

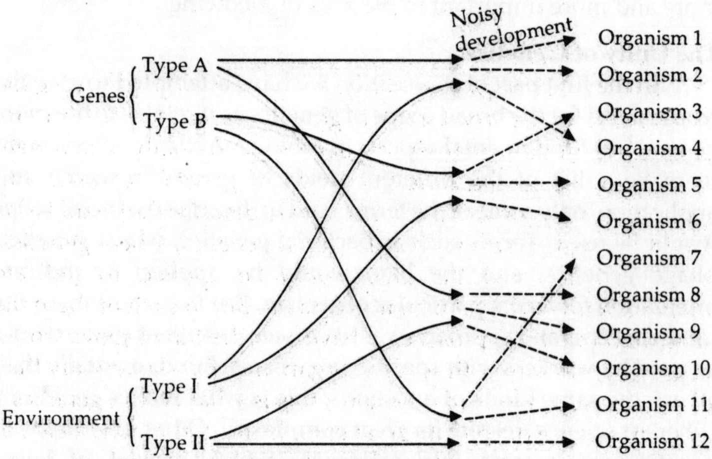

Fig. 1. A model of phenotypic determination that shows how genes, environment and developmental noise interact to produce any given phenotype.

The use and analysis of hereditary variants embodies the primary research tool of the geneticist. This approach is equally powerful from the molecular level, where it can be used to probe cellular and organismal processes, all the way to the population level, where it can be used to investigate evolutionary processes. The spectrum of variants that has been obtained and studied effectively by geneticists is staggering; as we have seen, there are variants affecting shape, number, biochemical function, and so on. In fact, the general finding has been that variants can be obtained for virtually any biological structure or process of interest to the investigator. Recall that a set of variants of any given biological character can fall into either of two broad categories: (1) those whose underlying genotypes show nonoverlapping norms of reaction, and (2) those whose underlying genotypes show overlapping norms of reaction. The techniques of genetic analysis are different in each case.

The analysis of genotypes with nonoverlapping norms of reaction has produced most of the phenomenal success of genetics in elucidating the cellular and molecular processes of organisms. The appeal of such systems from the experimenter's viewpoint is that the environment can virtually be ignored, because each genotype produces a discrete, identifiable phenotype. In fact, geneticists have deliberately sought and selected just such variants. Experiments become much easier—in fact, they only become possible—when the phenotype observed is an unambiguous indication of a particular genotype.

Mendel used genotypes that were established horticultural varieties. For example, in one set of variants, one genotype always produces tall Pea plants, and another always produces dwarf plants. Had Mendel chosen variants whose genotypes have overlapping norms of reaction, he would have obtained complex and patternless results like those in the *Achillea* experiments. He could never have identified the simple relationships that became the foundation of genetic understanding. Much of modern molecular genetics is based on research with variants of bacteria that similarly show distinctive characters whose genotypes have nonoverlapping norms of reaction. For instance, modern DNA manipulation technology relies heavily on selection systems based on bacterial genes for drug resistance; the expression of such genes is very clear-cut and reliable. In this category, the variants tend to be rather drastic. Many such variants could never survive in nature, simply because they are so developmentally extreme. In other cases, such as the winged and wingless fruits of *Plectritis*, the strikingly different forms do appear regularly in natural populations.

Reflecting the importance of discrete variants in developing our present understanding of the genetic basis of life, most of the rest of this book is devoted to analysis of this kind of variation.

The story begins with Mendel's research and theories, proceeds through classical genetics, and ends with the startling discoveries of molecular biology. It must be emphasised that the entire development of this knowledge critically depended on the availability of phenotypes having simple relationships to genotypes. Because our study of genetic analysis necessarily puts so much emphasis on such simple trait differences, you may get the impression that they represent the most common relationship between gene and trait. They do not.

Geneticists have to pick and choose among cases of genetic variation to find those with a simple correspondence between genotype and phenotype. For example, among the hundreds of classical mutants of *Drosophila* that have been well-enough characterised to place them on specific chromosomes, about half are too variable in their phenotypic expression to be used in further genetic analysis. Only a quarter are ideally suited for the purposes of careful genetic analysis. And this count excludes the large number of mutants that were so unreliable that they could not be put on genetic maps in the first place. Even among the most useful mutants, environment and developmental age can be quite important to their expression. The widely used mutation *purple* produces an eye colour like the normal ruby colour in very young flies, which darkens to a distinguishable difference only with age. The mutation *Curly*, one of the most important tools in genetic analysis in *Drosophila*, results in curled wings at 25°C, but not at 19°C, where the flies have normally straight wings.

Whereas simple one-to-one relationships between genotype and phenotype dominate the world of experimental genetics, in the natural world such relationships are quite rare. When norms of reaction for size, shape, colour, metabolic rate, reproductive rate, and behaviour have been obtained for organisms taken from natural populations, they almost always turn out to be like those of *Achillea*. The relationships of genotype to phenotype in nature are almost always one-to-many rather than one-to-one. This is the underlying cause of the rarity of discrete phenotypic classes in natural populations. Obviously, the analysis of these one-to-many relationships is far more complex. The researcher is confronted with a bewildering range of phenotypes. Special statistical techniques must be used to disentangle the genetic, environmental, and noise components.

The major discoveries of genetics were founded on work with discrete norms of reaction. Although these discoveries have revolutionarised pure and applied biology, great care must be taken in extrapolating such ideas to systems that are based on complex interaction of gene and environment, especially those found in nature. For example, the difference between yellow-bodied and gray-bodied *Drosophila* in laboratory stocks can be shown to be a simple genetic difference, but it does not follow that skin colour in humans obeys similar genetic laws. In fact, even in *Drosophila*, the various intensities of black pigment in flies from natural populations turn out to be a consequence of the interaction of temperature with a complex genetic system.

Genetics and Human Affairs

Knowledge about hereditary phenomena has been important to humans for a very long time. Civilisation itself became possible when nomadic tribes learned to domesticate plants and animals. Long before biology existed as a scientific discipline, people selected grains with higher yields and greater vigor and animals with better fur or meat. They also puzzled about the inheritance of desirable and undesirable traits in the human population. Despite this long-standing concern with heredity and the practice of selective breeding, it was not until the discovery of Mendel's laws that we were able to elucidate the actual basis for inheritance.

Early in this century a new wheat strain called Marquis was developed in Canada. This high-quality strain is resistant to disease; furthermore, it matures two weeks earlier than other commercially used strains—a very important factor where the growing season is short. At the time of its introduction, the use of Marquis wheat opened up millions of square miles of fertile soil to cultivation in such northern countries as Canada, Sweden, and the USSR. Table shows how geneticists have bred a wide range of desirable characteristics into another commercial crop, rice.

In addition to improving crop varieties, geneticists have learned to alter the genetic systems of insects to reduce their fertility. This technique is providing an important new weapon in the age-old struggle to keep insects out of human crops and habitations.

Such successes in recent years led to the concept of the "Green Revolution" as the scientific answer to the problem of human

Table 1
Development of Pest-resistant Strains of Rice

Strain	Year developed	Diseases					Insects		
		Blast fungus	Bacterial blight	Leaf streak virus	Grassy stunt virus	Tungro virus	Green leaf-hopper	Brown hopper	Stem borer
IR 8	1966	MR	S	S	S	S	R	S	MS
IR 5	1967	S	S	MS	S	S	R	S	S
IR 20	1969	MR	R	MR	S	R	R	S	MS
IR 22	1969	S	R	MS	S	S	S	S	S
IR 24	1971	S	S	MR	S	MR	R	S	S
IR 26	1973	MR	R	MR	MR	R	R	R	MR

Note: The entries describe each strain's susceptibility or resistance to each pest as follows: S = susceptible; MS = moderately susceptible; MR = moderately resistant; R = resistant.

hunger. Using sophisticated breeding techniques based on new knowledge about genes, geneticists created high-yield varieties of dwarf wheat and rice. Extensive planting of these crops around the world did provide new food supplies, but new problems quickly became apparent. These specialised crops require extensive cultivation and costly fertilisers. The use of the new high-yield varieties produced a wide range of social and economic problems in the impoverished countries where they were most needed. Furthermore, the spread of monoculture left vast areas at the mercy of some newly introduced or newly evolved form of pathogen—say, a plant disease or an insect pest. With the huge population of humans on earth, our dependence on high-yield varieties of crop plants and domestic animals is becoming increasingly obvious. In a very real sense, the stability of human society depends on the ability of geneticists to juggle the inherited traits that shape life forms, keeping the crops a jump ahead of the destructive parasites and predators. In recent years, advances in biotechnology have led to the creation of special genetically engineered strains of bacteria and fungi that carry specific genes from unrelated organisms such as humans. The microbes produce such useful compounds as insulin, human growth hormone, and the antiviral agent interferon.

The most exciting and frightening application of genetic knowledge is to the human species itself. Genetic discoveries have had major effects on medicine. We can now diagnose hereditary disease before or soon after birth, and in some cases we can provide secondary treatments. Using family pedigrees, a genetic counselor can give prospective parents the information they need to make intelligent decisions about the risks of genetic disease in their offspring. Such refined techniques as amniocentesis and fetoscopy provide information about possible genetic disease at early stages of pregnancy.

A battery of postnatal chemical tests can detect problems in the newborn infant, so that some corrective techniques can be applied immediately to alleviate the effects of many genetic diseases. Our new ability to recognise genetic disease poses an important moral dilemma. An estimated 5 per cent of our population survives with severe physical or mental genetic defects. This percentage probably will increase with extended exposure to various environmental factors—and, paradoxically, with improved medical technology. As geneticist Theodosius Dobzhansky has remarked,

> If we enable the weak and the deformed to live and propagate their kind, we face the prospect of a genetic twilight. But if we let them die or suffer when we can save or help them, we face the certainty of a moral twilight.

Of those patients admitted to pediatric hospitals in North America, 30 per cent are estimated to have diseases that can be traced to genetic causes. The financial burden to society is already significant.

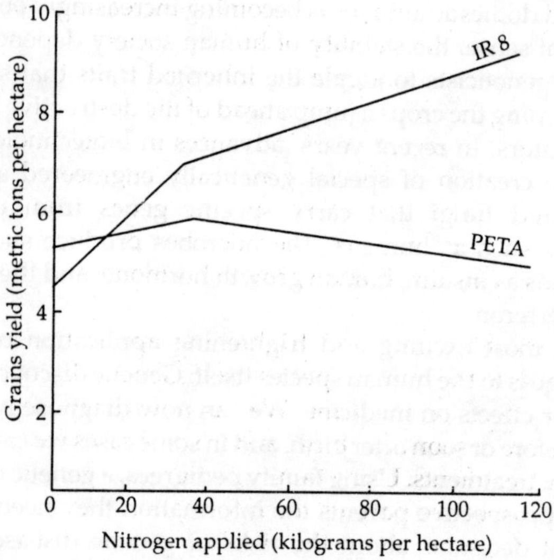

Fig. 2. The specially bred strain of dwarf rice called IR 8 owes part of its success to its remarkable response to the application of fertiliser. The strain PETA is an older, nondwarf strain showing a more typical response.

The human brain is subject to the same rules of genetic determination as the rest of the body. Does this mean that our thoughts and behaviour are extensively determined by inherited predispositions? Or can we view the mind as a clean slate at birth, written upon only by individual experience? The nature of inborn constraints on thought and personality—the implications for present sociological problems—have fascinated many geneticists and other scientists. Such books as *African Genesis, The Territorial Imperative, On Aggression,* and other popular titles on sociobiology have stimulated widespread public interest. A bitter debate has raged about the differences in intelligence among various racial and social groups.

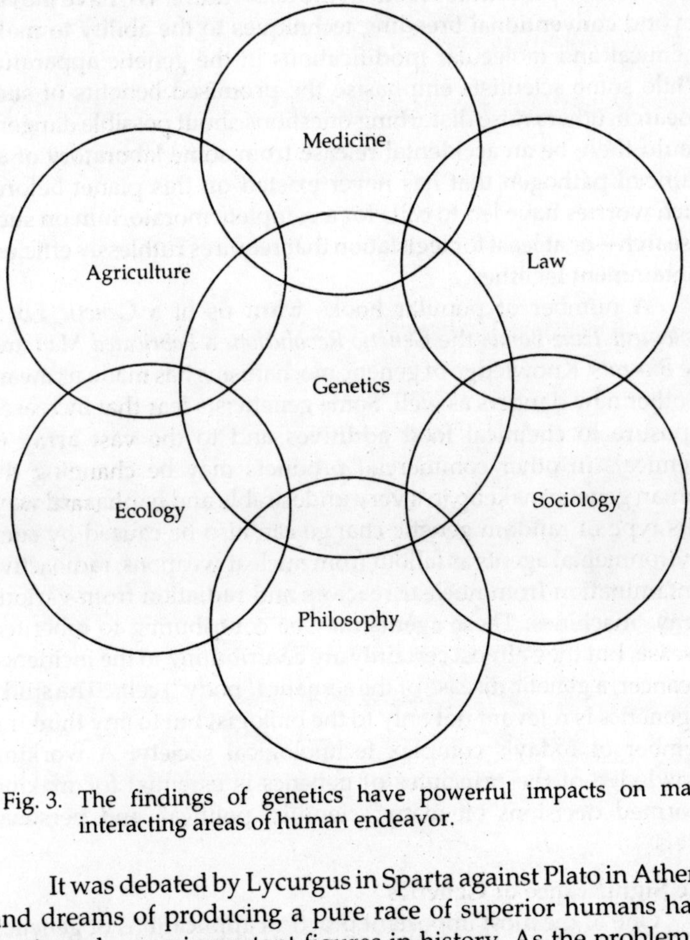

Fig. 3. The findings of genetics have powerful impacts on many interacting areas of human endeavor.

It was debated by Lycurgus in Sparta against Plato in Athens, and dreams of producing a pure race of superior humans have motivated many important figures in history. As the problem of human overpopulation becomes obvious to almost everyone, there is increased talk of legislated sterilisation and the planned selection of human offspring. There is very serious talk, even among geneticists, about the ability of the human race to take control of its own evolution. Others are frightened by the possibilities for disastrous error or unpleasant sociological consequences. The sophisticated technology of molecular genetics has given us a wide range of new techniques for shaping our genetic makeup. Even more bizarre procedures loom in the near future. We have moved beyond conventional breeding techniques to the ability to make chemical and molecular modifications in the genetic apparatus. While some scientists emphasise the promised benefits of such research, others raise disturbing questions about possible dangers. Could there be an accidental release from some laboratory of an artificial pathogen that has never existed on this planet before? Such worries have led to calls for a complete moratorium on such research—or at least for legislation that requires ruthlessly efficient containment facilities.

A number of popular books warn us of a *Genetic Fix, a Biological Time Bomb,* the *Genetic Revolution,* a *Fabricated Man* and the *Biocrats.* Knowledge of genetic mechanisms has made us aware of other new dangers as well. Some geneticists fear that increased exposure to chemical food additives and to the vast array of chemicals in other commercial products may be changing the human genetic makeup in a very undesirable and haphazard way. This type of random genetic change can also be caused by such environmental agents as fallout from nuclear weapons, radioactive contamination from nuclear reactors and radiation from various X-ray machines. These agents may be contributing to inherited disease, but they almost certainly are contributing to the incidence of cancer, a genetic disease of the somatic ("body") cells. The study of genetics is relevant not only to the biologist but to any thinking member of today's complex technological society. A working knowledge of the principles of genetics is essential for making informed decisions on many scientific, political, and personal levels.

The Significance of Genetics

One of the most important practical applications of genetics is in plant and animal breeding. Practical breeders draw on the immense store of basic genetic knowledge and use it to develop economically superior varieties and breeds of plants and animals to help feed the hungry of the world.

Breeding

The continuity of life from generation to generation was certainly recognised for the various plant and animal forms, although hardly understood until the nineteenth century. At some time in the past, humans began, perhaps first unconsciously and then consciously, to domesticate certain animals and selectively breed them to fit them to their needs. Our present-day cattle, sheep, goats, horses, cats and dogs are descended from these early beginnings of breeding, started perhaps as long as 10,000 to 15,000 years ago in the Stone Age. The success of the selective breeding programmes is amply illustrated by the wide variety of different breeds of dogs and horses adopted to carry out various kinds of functions. Most of these were developed long before the nineteenth century. It was not until humans began to cultivate plants, however, and abandoned their nomadic existence, associated with the herding of herbivorous animals, that the really important applications of genetics began. About 6000 to 7000 years ago, the cultivation of plants that became our present-day staples began. Plants such as yams, rice, wheat, barley, oats, corn, and beans were derived from the wild forms and grown in fields. These became the mainstays of the human diet, more important even than the flesh of animals.

Over the many-thousand-year period up to the last century, plant breeders produced many varieties of plants that served as food or as ornamentals. All this early breeding work had the status of an art. It was done empirically but was, in fact, highly successful, and it formed one of the bases for the development of cities and civilisation. With the rediscovery of Mendelism in 1900, there was an almost immediate application of Mendelian principles to plant and animal breeding. For the first time breeding was put on a scientific basis. It can be stated unequivocally that production of food from every important food plant and animal has benefitted and been increased by the application of genetics.

Hybrid Breeding in Plants

In 1909, G.H. Shull, the geneticist, discovered that he could produce a superior strain of corn by hybridizing two *inbred* strains. By inbred we mean lines in which plants had been self-pollinated for many generations. From these early beginnings, and with the further work of many other corn geneticists, the present-day hybrid corn has been developed, and is now virtually the only kind of corn grown in the United States. Prior to the use of the hybrid strains of corn, production was on the order of 30 bushels per acre in the best corn-producing areas. Average yields are now 90 to 100 bushels per acre, a threefold increase! Significant increases in the yield and food quality of other staple food plants such as sorghum, sugar beets, onions, tomatoes, and carrots have also been achieved by the use of hybridizing techniques similar to those used in corn. Not all of this increase can be attributed to genetics alone, however. The use of fertilisers, insecticides, and herbicides has also been a large factor increasing the yield of all important food plants.

Other Kinds of Plant Breeding

The increase in agricultural production is not only the result of producing strains that are able to grow faster and produce more by virtue of their efficiency in converting sun energy into food because of a superior metabolism. It is also the result of utilising genetic techniques to produce strains of plants that are resistant to animal pests such as insects and nematodes, and to various kinds of pathogenic microorganisms that can wreak havoc with crop plants, particularly in the breadbasket areas where many thousands of contiguous acres may be planted with the same kind of plant. Plants can also be bred to be adaptable to mechanical harvesting. This has been done even with as dubious a candidate as the tomato. The ordinary tomato bought fresh in the vegetable market obviously would create a mess if picked mechanically, but varieties have been developed with tough skins that are easily picked by machines. When the skins are removed at processing plants, the remains constitute the present-day major source of canned tomatoes, tomato paste, and tomato juice.

Plant geneticists have also been highly successful in developing new crop and ornamental plant varieties by crossing together two different species to produce a more desirable interspecific hybrid, or by merely introducing a small but desirable part of the chromosome complement of one species into an already established crop plant of another species. For example, the cultivated wheat plant, *Triticum aestivum,* is the western world's major food staple, equalled in importance only by the rice plant, *Oryza sativa,* in the Orient. Like most plants, cultivated wheat has little or no resistance to certain diseases that can cause great reductions in crop yield. Other relatives of wheat do show resistance to these diseases; however, it is not feasible to make complete hybrids between the cultivated wheat and these resistant relatives, because the hybrids, although they may show some resistance, do not produce good wheat flour. Therefore, procedures have been worked out so that by "chromosome engineering" only those small parts of the chromosomes of the resistant plant that bear the genetic material that confer resistance are transferred to cultivated wheat.

The activities of plant geneticists described above have in part been responsible for what is called the "Green Revolution." There has been a dramatic increase in food production in the world during the last 20 years. This is not the result of increasing the acreage planted, but of using new varieties of plants and developing more efficient farming practices. The breeding of new varieties of plants must be a continuing process; new resistant strains must be developed because old ones may no longer be able to cope with the newly acquired virulence of pests or new climatic conditions.

Animal Breeding

Since the application of genetic knowledge to the production of food from food animals, there have been enormous increases in total animal food produced per ton of plant food ingested by the animals. Cows now give more milk and tenderer beef; chickens produce more eggs, and the chicks grow faster to produce more meat per pound of grain eaten; turkeys have more white meat; bees produce more honey because they work harder; pigs produce more and leaner pork, and so on. Much of the success in the breeding of better food animals has been the result of the widespread use of artificial insemination, particularly among mammals. You may have wondered why at stock shows prize bulls may go for prices up to and over several hundred thousand dollars. Surely no single animal has been worth that much! The fact is the bull is valuable because of his sperm. His build and lineage make it highly probable that he will produce superior progeny (provided he is not sterile, which sometimes happens with these carefully bred animals). If his sperm are viable, they can be frozen and sent all over the world. A single bull can in this way be the father of thousands of offspring, and the sale of his sperm will more than pay for his original price and subsequent care.

Biological Control

Various kinds of insects and other animals, and the fungi, bacteria, and protozoa have plagued agriculture as long as food plants and animals have been raised. Over the years various devices have been resorted to in order to combat these pests, including the development of resistant plant strains by selective breeding as already mentioned. While this application of genetics has been effective to a degree, it still has rather serious drawbacks because the predators and pathogens may change as fast as the geneticist can develop plants resistant to them. In addition, it is not always possible to find plant strains that are resistant to all the different kinds of pests. For this reason, it has continued to be necessary to use enormous quantities of insecticides, fungicides, and other pesticides to keep plant and animal production at profitable levels. As everyone knows, the use of these chemicals, most of which are not biodegradable, presents the serious problem of contamination of the environment. These chemicals are not specific; they can poison other forms of life—including humans.

One method has been used with moderate success to control the screwworm, the larva of the fly, *Callitroga hominivorax*, which is a serious pest to livestock in the southern part of the United States and Mexico. These flies are raised in large numbers in special laboratories, and the adults are sterilised by use of X-rays. The sterile males are then released in large numbers into the air.

The screwworm fly normally maintains a low population density. Therefore screwworm populations are readily saturated with sterile males. The result is that only infertile eggs are produced. Other insects that are pests of both plants and animals can probably be similarly controlled without real damage to the environment. Another possible means of controlling insects is to take advantage of the fact that they usually have complex mating patterns that involve chemical attractants called *pheromones*. Mutant insect strains producing the wrong (ineffective) pheromones could theoretically be raised in large numbers and distributed in those areas where their nonmutant form is extant. This dilution of the wild population by the mutant forms might well cause a breakdown in the mating of males and females and a large reduction in the size of the next generation. These are but examples of possibilities for the control of pests. None of them, however, can be brought to fruition without the application of genetic knowledge.

Medicine and Society

It has not been until relatively recently (since World War II) that medical researchers and practitioners have begun to pay much attention to genetics as an aspect of medicine. There are probably two reasons for this long neglect of genetics on the part of the medical profession. The first was a general abhorrence of the ideas of the eugenicists. The eugenics movement arose in a formal way in the latter part of the nineteenth century and persists in one form or another today. In general, the extreme form of eugenics maintains that human beings should be bred as are animals for the production, in future generations, of superior men and women. It has led to laws requiring the sterilisation of individuals considered inferior. In its most extreme form, it led Adolf Hitler to order the murder of millions of Jews in Europe during World War II. The second reason medicine tended to ignore genetics was the idea that, although there may exist diseases that are inherited, there was nothing medicine could do about them. A genetically determined disease followed an inevitable course that could not be modified. We know now that this is not always true.

Today many members of the medical profession recognise that genetics can be a powerful ally, and nearly all medical schools have departments of genetics or at least geneticists on their faculties. This change in attitude has, in large part, been due to the great advances made in the last 20 to 30 years in our uderstanding of human metabolism and the realisation that it is under the control of genes. Also, the discoveries in human cytogenetics made in the last 15 to 20 years have opened the eyes of physicians to the importance of genes and chromosomes.

To give but a few examples, it may be pointed out first that of the half million or so spontaneous abortions occurring among pregnant women each year in the United States at least half are almost certainly caused by severe errors in the chromosome complements of the embryos. Second, a large fraction of the children born mentally retarded have genetic abnormalities. Third, probably 20 per cent of the children admitted to pediatric clinics in the major medical centres have genetic diseases, and 50 percent more have diseases that are genetically mediated (i.e., they appear to be inherited but not in any simple pattern). Fourth, it is now well demonstrated that the course of several genetic diseases can be altered by proper treatment, and it is almost certain that this number will increase as more is learned.

Much can be done about the occurrence of the 1800 or more genetic diseases known to exist in the human population by the use of genetic counselling. An understanding of these diseases by the physician and by the geneticist can certainly help alleviate the effects of these in individuals in which they occur and in the human population as a whole. More recent than the wide application of genetic knowledge to the understanding of human disease has been the realisation by psychologists, psychiatrists, and behaviourists in general that there is a genetic component to behaviour in human beings and in other animals.

1

Mendelian Principles

Multiplication is the unique feature of living organisms. All organisms arise only from preexisting living individual organisms. Thus, every individual organism must be viewed as the current member of an uninterrupted line of successive generations leading back to the origin of its species and ultimately to the dawn of life itself. This chapter discusses the basic principles of *genetics,* the science of *heredity*—the transmission of characteristics from one generation to the next—and *variation*—the changes that frequently occur in heritable characteristics. It has been observed from ancient times that the offspring of all organisms resemble their parents.

The essence of heredity is this recurrent appearance in successive generations of the characteristics of a species, defined as one or more populations of an organism that interbreed but do not in nature exchange genes with other organisms. Besides expressing the general features of its species, however, each organism exhibits individual traits. Many of these can also be traced from one generation to the next, since they too are governed by the laws of heredity.

Fundamentals of Inheritance

When Mendel began the experiments that would make him famous, almost nothing was known of the physical basis of inheritance. Concepts such as haploidy, diploidy, and meiosis were unknown at the time, and it had not even been established that nuclei carried genetic information. Understanding the physical basis of inheritance was to await the further development of the light microscope and staining techniques near the close of the century. However, many biologists, among them Charles Darwin, had been *hybridizing,* that is, mating different strains of plants and animals in attempts to gain insight into the mechanisms of inheritance and, more important, into the basis for the formation of new species. It was known at this time that in sexually reproducing organisms, both parents might contribute traits to their offspring. Moreover, in crosses between dissimilar parents, it has been noted that some of the traits of one parent might be expressed to the exclusion of the alternative traits of the other. When hybrids are interbred, more parental forms are produced.

Innovations of Mendel

Mendel's innovative approach to studying inheritance included the following considerations:

1. Mendel deliberately searched for a plant species (a) the varieties of which differed from each other in constant, distinct and mutually exclusive ways and lacked intermediate varietal types; (b) whose flowers normally underwent self-pollination and were naturally protected from chance pollination from another plant; and (c) that would continue to breed with full fertility for many generations after being hybridized. He finally selected on the common garden pea *(Pisum sativum)* as possessing the necessary qualifications.
2. When Mendel crossed different strains of pea plants, he followed the inheritance of only one set of contrasting traits at a time, ignoring the inheritance of all other traits.
3. Mendel counted and recorded the thousands of progeny from his experiments and then analyzed his results statistically.

Mendel chose two stocks with contrasting traits to perform an experiment. He then opened the immature flowers of the strain he chose to be the seed-bearing parent and removed all its anthers, the expanded tips of stamens containing pollen, before they produced any pollen. Then he dusted pollen from the other strain, the pollen-bearing parent, on the stigma of the seed-bearing parent and covered the flower to prevent further pollination. To cross two members of the same genetic line, he repeated this process with two similar individuals; alternatively, he permitted *selfing* to occur by allowing the anthers of a flower to mature and pollinate the carpels (where the ovules are found) of the same flower, thus, in effect, crossing two identical individuals.

Mendelism

Gregor Mendel is often called the father of genetics, and he postulated the basic laws of inheritance. One factor that may have contributed to Mendel's discoveries was his experience with plants during his youth on a farm. He also carried on plant breeding experiments that led to the development of prize-winning new varieties of fruits and vegetables. As a young, largely self-taught scientist, he joined a teaching order of Augustinian monks near Brünn, Austria (now Brno, Czechoslovakia), partly because it was the only way for him to pursue a scholarly career. He taught science in local secondary schools for a time before being sent to the University of Vienna for formal study in botany, physics and applied mathematics. Upon returning to Brünn, he established 34 pure-breeding lines of garden peas on the monastery grounds in order to experiment with them. (Pure-breeding lines always produce similar offspring.)

Plants in Heredity

Plants are easily grown in large quantities, they often produce large numbers of offspring, and many have relatively short generation times. In many plant species, the same individuals have both male and female reproductive organs, in which case each plant may reproduce as a male, as a female, or as both. Best of all, it is often easy to control which individuals mate. Some observations that Mendel found useful in his studies had been made in the late eighteenth century by a German botanist, Josef Gottlieb Kölreuter. Kölreuter attempted many crosses between plants, producing many *hybrids* and learning a great deal about the biology of plant reproduction. He was the first to demonstrate that the two parents play equal roles in heredity, and he studied the offspring from *reciprocal crosses*—matings made in two directions. For example, in one set of crosses, males that have white flowers are crossed with females that have red flowers, while in a complementary set of crosses red-flowered males and white-flowered females are the parents. Such crosses are called *reciprocal crosses* and always gave identical results. Laws of inheritance proposed at the time favoured the concept of *blending*. If a plant that had one form of a characteristic (say, red flowers) were crossed with one that had a different form of that characteristic (blue flowers), the offspring would be a blended combination of the two parents (purple flowers). And following the blending motif, it was thought that once heritable elements were combined, they could not be separated again (like combined inks). The red and blue genetic determinants were thought to be forever blended into the new purple one. Then, about a century after Kölreuter completed his work, Mendel began his.

Mendel's Discoveries

When Mendel began his research, it was known that one female gamete combines with one male gamete to bring about *fertilisation*. However, the role of the chromosomes as bearers of genetic information was unknown, and mitosis and meiosis were yet to be discovered. Mendel himself was well-qualified to make the big step forward. Although in 1850 he had failed an examination for a teaching certificate in natural science, he later undertook intensive studies in physics, chemistry, mathematics, and various aspects of biology at the University of Vienna. His work in physics and mathematics probably led to his applying experimental and quantitative methods to the study of heredity—and these were the key ingredients in his success. Mendel worked out the basic principles of the heredity of plants and animals over a period of about nine years, the work culminating in a public lecture in 1865 and a detailed written account published in 1866. Mendel's paper on plant hybridisation appeared in a journal that was received by 120 libraries, and he sent reprinted copies (of which he had obtained 40) to several distinguished scholars. However, his theory was not accepted. In fact, it was ignored. Perhaps the chief difficulty was that the physical basis of his theory was not understood until the discovery of meiosis, some years later.

The most prominent biologists of Mendel's time were not in the habit of thinking in mathematical terms, even the simple terms used by Mendel. Whatever, the reasons, Mendel's pioneering paper had no discernible influence on the scientific world for more than 30 years. Then, in 1900, Mendel's discoveries burst into prominence as a result of independent experiments by the Dutch Hugo de Vries, the German Karl Correns, and the Austrian Erich von Tschermak. Each of these scientists carried out crossing experiments and obtained quantitative data about the progeny; each published his principal findings in 1900; each cited Mendel's 1866 paper. At last the time was ripe for biologists to appreciate the significance of what these four geneticists had discovered—largely because meiosis had by then been described. That Mendel was able to make his discoveries *before* the discovery of meiosis was due in part to the methods of experimentation he used.

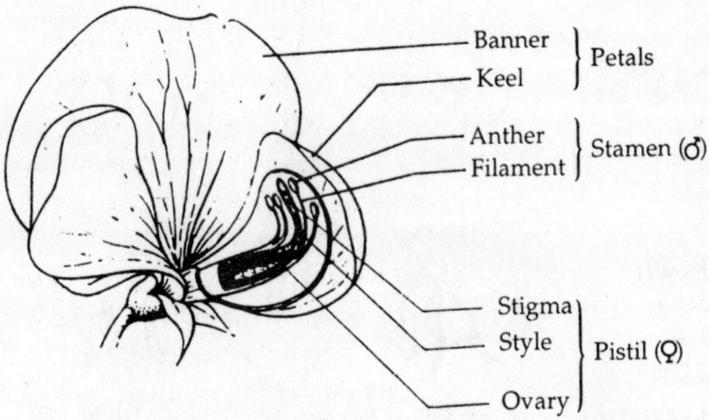

Fig. 1. A pea flower from which a section has been removed to show the reproductive structures. Each flower has a single ovary, containing up to ten ovules, which develops into the seed pod. The pistil is shown in red.

Selection of a Genetic Material

Mendel selected peas for his experiments because they possessed many desirable features:

1. Numerous varieties of peas were available commercially. They provided the material that he studied.
2. The plants were easy to cultivate and the generation time was short.
3. The offspring of the crosses between the varieties were fertile.
4. The structure of the pea flower is such that accidental pollination was thought not to occur. The anthers that produce the pollen and the stigma where the pollen grains germinate, are completely enclosed by the petals.

Self-fertilisation pollen from a flower falls on the stigma of the same flower. In those cases where crosses between varieties are desired, it is possible to remove the anthers before they mature and somewhat later, when the stigma is mature, to cover it with pollen from another flower.

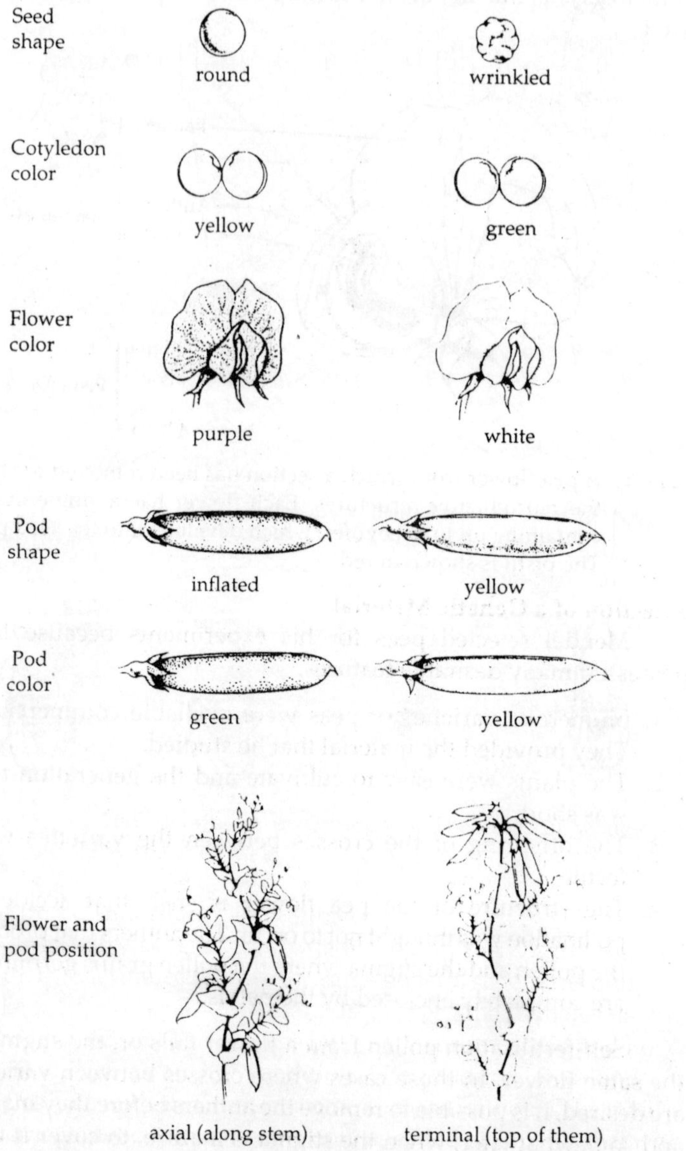

Fig. 2. Six of the seven character differences in peas studied by Mendel (the seventh difference was long versus short stems). In each case the characteristic on the left is seen in a hybrid.

Mendel selected seven pairs of contrasting characters. Some of his varieties had *round* seeds; others had *wrinkled* seeds. In all he studied seven different characters of the pea, and for each he had two varieties, as shown in the following lists:

Character Affected	*Varieties*
Seed shape	*round or wrinkled*
Seed colour	*yellow* or *green*
Seed coat colour	*coloured* or *white*
Pod shape	*inflated* or *wrinkled*
Pod colour	*green* or *yellow*
Flower position	*axial* or *terminal*
Stem length	*long* or *short*

Crosses Made by Mendel

First, Mendel made sure that all of his varieties would breed true. Once this was established he made crosses between all of the pairs just listed. The results were most unexpected in the light of earlier experiments by other plant breeders. The offspring were never intermediate but were always like one of the parents. When peas with *round* seeds were crossed with peas with *wrinkled* seeds, for example, the offspring were plants with *round* seeds. Mendel spoke of the form that appeared in the offspring as *dominant* in comparison to the form that did not appear, which he called *recessive*. The varieties that Mendel used were found to have these relationships:

Dominant	*Recessive*
round seed	*wrinkled* seed
yellow seed	*greed* seed
coloured seed coat	*white* seed coat
inflated pod	*wrinkled* pod
greed pod	*yellow* pod
axial flowers	*terminal* flowers
long stem	*short* stem

Over the course of years, geneticists have introduced some terms that make it easier to discuss crosses. The original parental generation is abbreviated P. The offspring of the *P* generation is the first filial or F_1 generation. The offspring of the F_1 is the second filial generation or F_2, the third is the F_3, and so on. It is also customary to describe crosses in terms of the character. Thus, the cross of a plant with round seeds with a plant with wrinkled seeds is shortened to *round × wrinkled.*

Second Filial Generation

Some plant breeders might have stopped the experiments after a single cross had determined dominant and recessive characteristics. The results, after all, were uniform and clear cut. The F_1 plants were always like one of the parents. Mendel, however, continued his crosses and was careful to realise that although an F_1 of the *round × wrinkled* might be identical with the *round* parent, its parentage was different. Perhaps its genetic behaviour would reflect the different origin. Since peas are self-fertilizing, the F_1 plants pollinated their own ovules and gave the F_2, and when Mendel studied the F_2 plants he found that both dominant and recessive characters were present. Mendel did a simple though revolutionary thing: he counted the number of individuals of each type. In every cross there was a ratio of 3 dominant to 1 recessive. His results can be summarised as follows:

P	F_1	F_2 *Counts*	*Ratio*
round × wrinkled	round	5,474 round	
		1,850 wrinkled	2.96 : 1
yellow × green	yellow	6,022 yellow	
		2,001 green	3.01 : 1
coloured × white	coloured	705 coloured	
		224 white	3.15 : 1
inflated × wrinkled	inflated	882 inflated	
		299 wrinkled	2.95 : 1
green pods × yellow pods	green	428 green	
		152 yellow	2.82 : 1
axial × terminal	axial	651 axial	
		207 terminal	3.14 : 1
long × short	long	787 long	
		277 short	2.84 : 1

These results would suggest that the rules of inheritance were the same, irrespective of the varieties being crossed. The F_1 plants were always of one type, which resembled one of the parents. In the F_2, two classes appeared and the frequency was 75 per cent dominants and 25 per cent recessives, or a ratio of 3 : 1.

Third Filial Generation

Mendel continued his experiments and obtained an F_3 generation. We may take as an example of his work the *round* × *wrinkled*, which in the F_2 gave 75 per cent *round* and 25 per cent *wrinkled*. He allowed a number of the *wrinkled* plants to self-fertilise to give an F_3, and found that all bred true, that is, only *wrinkled* plants were obtained in the F_3. The F_2 *round* plants gave two results:

1. One-third (193 of 565 plants) bred true, giving *round* plants in the F_3.
2. Two-thirds (372 of 565 plants) gave *round* and *wrinkled* in a ratio of 3 : 1.

Table 1
The Seven Characteristics used by Gregor Mendel in his Studies of the Garden Pea

Seeds	Round *vs.* wrinkled
Seeds	Yellow *vs.* green
Flowers	Purple *vs.* white
Flowers	Axial *vs.* terminal
Pods	Inflated *vs.* pinched
Pods	Green *vs.* yellow
Stem	Tall *vs.* short

Law of Segregation

In interpreting these results, Mendel stated what is now called Mendel's *law of segregation*. Mendel concluded that characteristics such as tall or short stems are determined by discrete factors. We now call these factors *genes*. Mendel stated that *each organism contains two factors for each characteristic and that the factors segregate during the formation of gametes so that each gamete contains only one of each pair of factors.* Mendel was able to draw these conclusions from his experimental results even though he was unaware of the fact that his factors were located on chromosomes. He was also unaware of meiosis, which is, responsible for the segregation of chromosomes (and therefore genes) before the formation of gametes.

P_1 (first parental) Tall × Short

↓

F_1 Tall (all plants)

↓ Selfed

F_2 Tall (74%) Short (26 %)

Fig. 3. A diagram of Mendel's experiments on the inheritance of stem length in pea plants.

Figure given above shows how Mendel's theory can be used to interpret the results of his experiment on stem height in peas. Each pea plant has two copies of the gene that controls the height of the stem. In this example, each copy may be either of two possible forms. Alternate forms of a gene are called **alleles**. Thus, in this example there is an allele for tallness and an allele for shortness. The original parents used in Mendel's cross, referred to as the P_1 generation, were "true-breeding." Therefore, the tall plants had two copies of the allele for tallness and the short plants had two copies of the allele for shortness. When an organism has two of the same kind of alleles, it is *homozygous* for that gene. Because the first parents were homozygous, all gametes produced by the short plant contain the allele for shortness, and all gametes produced by the tall plant contain the allele for tallness. In the resulting F_1 generation, all the individuals have one allele for tallness and one for shortness. When an organism has two different alleles of a gene, it is **heterozygous** for that gene. Although the plants of the F_1 generation have one of each type of allele, they are all tall.

When expression of only one allele in a heterozygous individual is observed, this allele is a **dominant** allele. The allele whose expression is masked in a heterozygote is a **recessive** allele. In modern genetic notation, the dominant allele is indicated with an uppercase letter and the recessive allele with a lowercase letter (or, in some cases, + and –). For example, the allele for tallness in peas can be symbolized *T*, while the allele for shortness can be symbolized *t*.

Genotype and Phenotype

Two organisms with different allele combinations for a trait may have the same outward appearance (*TT* and *Tt* peas are both tall). For this reason it is important for us to distinguish between the genetic constitution and the outward appearance of an organism. The genetic constitution of an organism is its *genotype*.

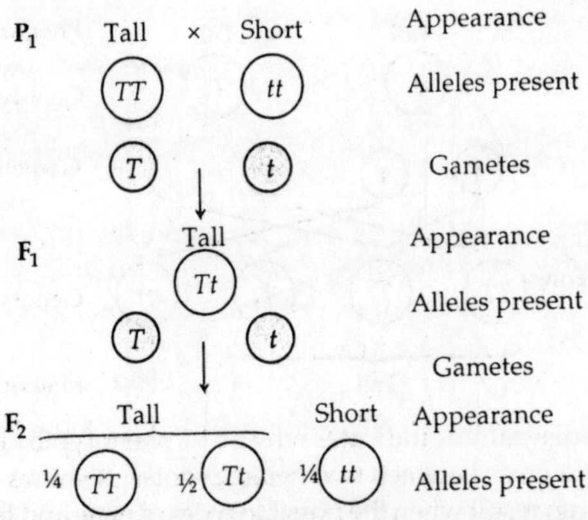

Fig. 4. Explanation of the genetic basis for the results of Mendel's experiment on stem length. Genetic makeup of all plants and gametes produced are shown.

The genotype is determined at fertilisation and does not change during the lifetime of the organism. The *phenotype* of an organism is a set of observable characteristics of that organism. In pea, both *TT* and *Tt* are genotypes that give the tall phenotype.

When heterozygous F_1 pea plants are selfed, the results are the same as those that would be obtained if two heterozygous individuals (heterozygotes) are crossed. Approximately half of the gametes (both male and female) that are produced have the allele for tall stems and half have the allele for short stems. When these combine to form F_2 zygotes, the results are one-fourth homozygous tall plants, one-fourth homozygous short plants, and one-half heterozygotes. Because the heterozygotes display the tall phenotype, 75 per cent of the F_2 generation are tall and 25 per cent are short, thus producing the 3 : 1 phenotypic ratio.

Law of Independent Assortment

Mendel studied seven pairs of characters in peas, involving seed colour, seed surface, flower colour, vine height, colour of unripe pod, pod shape, and position of flowers. One of his crosses was between a pea plant with round and yellow seeds and one with wrinkled and green ones. Such a cross, which involves two character differences separable in inheritance, is called a *dihybrid*

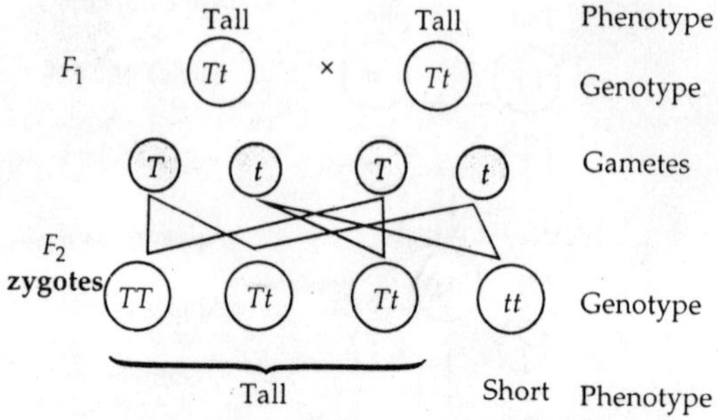

Fig. 5. This diagramme indicates why a 3 : 1 phenotypic ratio is expected from a cross between two heterozygotes. It shows the genotypes that can result when the possible types of male and female gametes combine.

cross (a cross involving a single pair of alleles is *monohybrid*). The F_1 hybrids all had yellow and round seeds, since the yellow colour is dominant over the green and the round shape is dominant over the wrinkled. When the F_1-hybrid plants were crossed to each other or allowed to self-fertilize, an F_2 generation was obtained, consisting of 556 seeds. Among these seeds the expected segregation in the ratio of 3 dominants : 1 recessive (¾ : ¼) was observed both for seed colour and for seed shape are given on next page.

416, or 74.82 per cent, yellow seeds
140, or 25.18 per cent, green seeds
423, or 76.08 per cent, round seeds
133, or 23.92 per cent, wrinkled seeds

However, the question is this: Is the segregation of the seed colour characters independent of that of the seed shape, or are the two characters somehow tied together? Since the original varieties crossed had yellow round and green wrinkled seeds respectively, will the yellow colour continue to be associated with the round shape and the green colour with the wrinkled shape of the seeds? Or will all combinations of the characters appear? Mendel found that the segregation of the seed colour is independent of the seed shape and that both the old (parental) and new combinations of the characters appear in the F_2 offspring. Classifying the 556 seeds for both characters at once (cf. Appendix), he obtained:

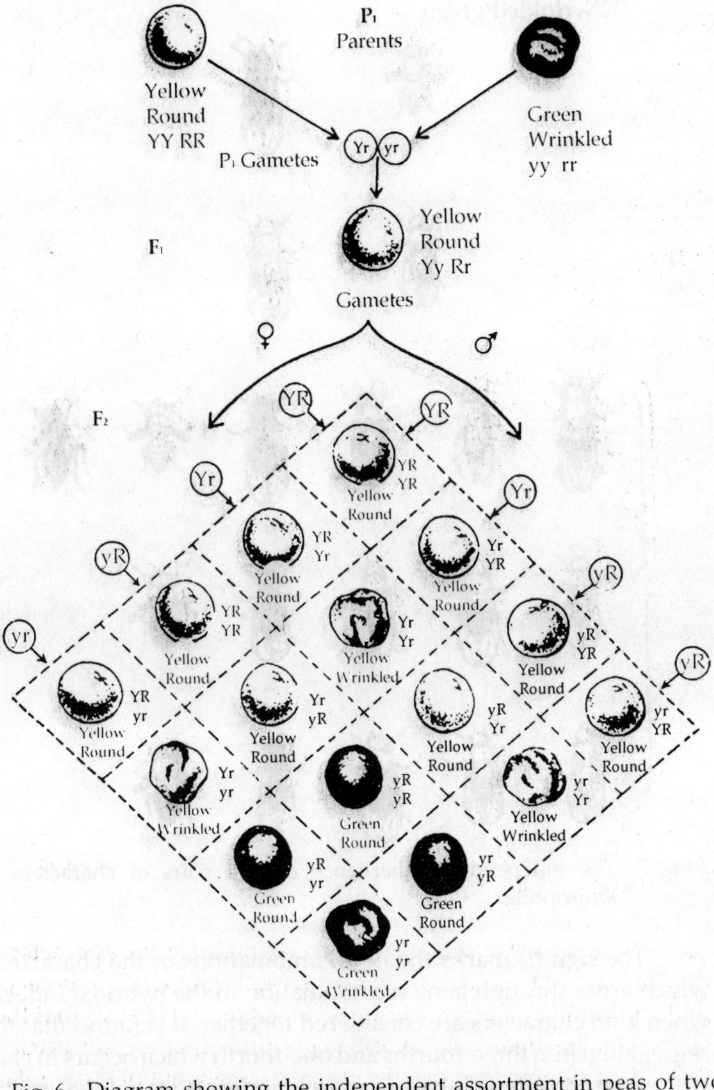

Fig. 6. Diagram showing the independent assortment in peas of two pairs of characters in which dominance is complete.

315 round yellow seeds
108 round green seeds (!)
101 wrinkled yellow (!)
32 wrinkled green

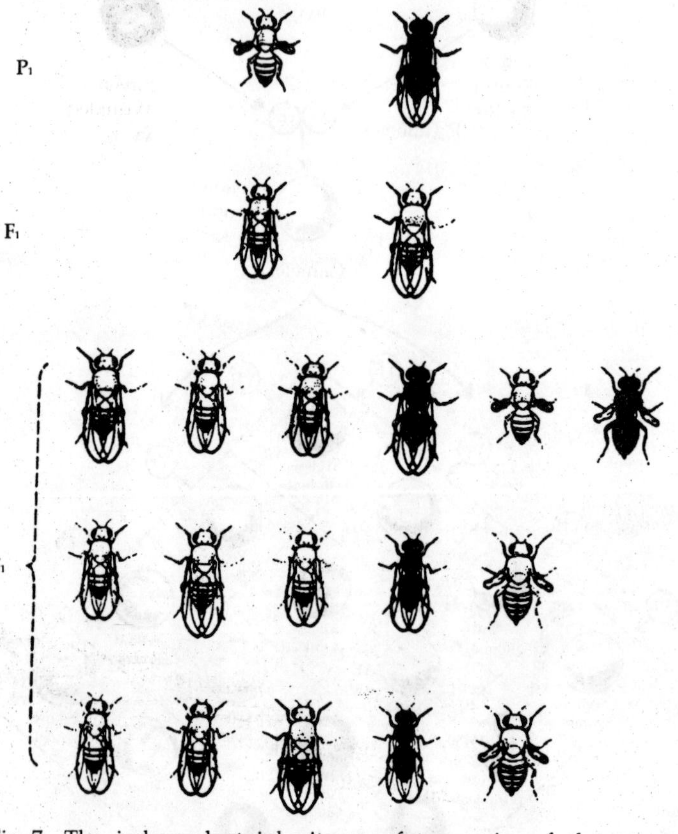

Fig. 7. The independent inheritance of two pairs of characters in Drosophila.

The sign (!) marks the new combinations of the characters, which arose through gene recombination in the hybrids. Indeed, when both characters are considered together, it is found that the segregation into three-fourths and one-fourth which occurs in each pair when considered alone is entirely independent of the similar segregation that takes place in the other pair. Thus, of the three-fourths of the entire group of plants which are round-seeded, approximately three-fourths, in turn, are yellow-seeded, and one-

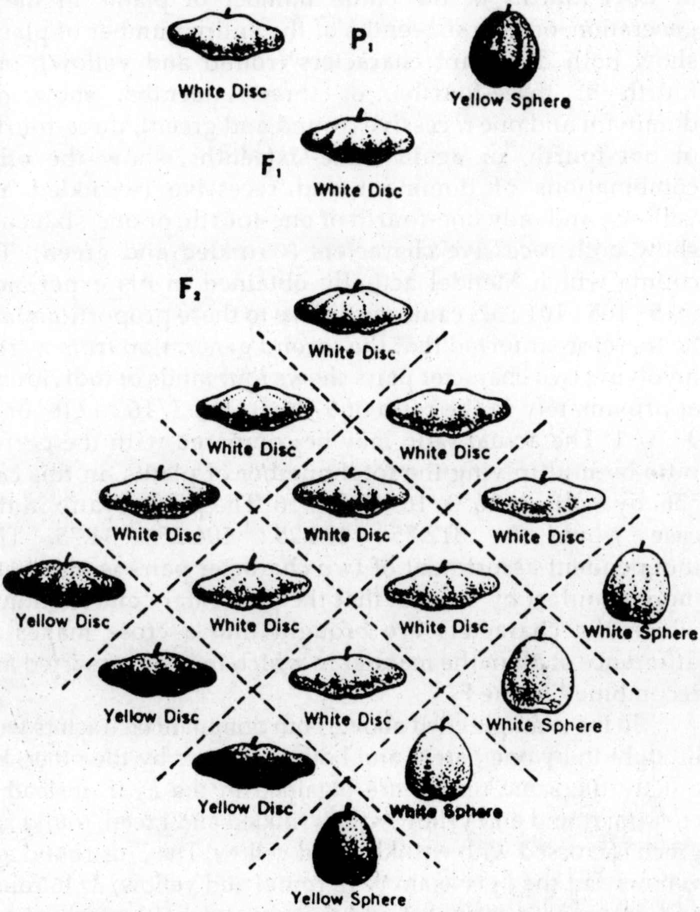

Fig. 8. The inheritance of two pairs of characters in summer squashes, illustrating Mendel's law of independent assortment. White is dominant over yellow and "disk" shape over "sphere". In the F_2 there result 9/16 white disk plants; 3/6 white sphere, 3/16 yellow disk; and 1/16 yellow sphere.

fourth are green; and of the other fraction (one fourth) which are wrinkled-seeded, three-fourths, again, are yellow-seeded and one-fourth green. This leads to the result that three-fourths of three-fourths of the entire number of plants in the F_2 generation, or nine sixteenths of the entire number of plants, show both dominant characters (round and yellow); one-fourth of three-fourths, or three-sixteenths, show one dominant and one recessive (round and green); three-fourths of one-fourth, or again three-sixteenths, show the other combinations of dominant and recessive (wrinkled and yellow); and only one-fourth of one-fourth, or one-sixteenth, show both recessive characters (wrinkled and green). The counts which Mendel actually obtained in his experiment (315 : 108 : 101 : 32) came very close to these proportions, and he therefore inferred that the second generation from a cross involving two character pairs shows four kinds of individuals, approximately in the ratio of 9/16: 3/16 : 3/16 : 1/16, or 9 : 3 : 3 : 1. The actual ratio may be compared with the perfect ratio by multiplying the total number of plants, in this case 556, by 9/16, 3/16, 3/16, and 1/16. The perfect ratio in this case would be 312.75 : 104.25 : 104.25 : 34.75. This independent assortment of two character pairs is made still more manifest by the fact that the particular combination in which the characters are brought into a cross makes no difference at all in the manner in which they are assorted and recombined in the F_2.

In the example cited above, both dominant characters were brought in by one parent and both recessives by the other, but exactly the same results are obtained in the F_2 if, instead of crossing round and yellow with wrinkled and green, round and green is crossed with wrinkled and yellow. The F_1 is round and yellow; and the F_2 is again 9/16 round and yellow; 3/16 round and green; 3/16 wrinkled and yellow; and 1/16 wrinkled and green.

Independent Assortment

Let the gene for the round seeds be represented by *R* and its allele for wrinkled seeds by *r;* and the gene for yellow seeds by *Y* and for green seeds by *y*. Mendel's original yellow round-seeded parent thus had the genes *RRYY,* and the green wrinkled parent *rryy*. The gametes produced by these parents were respectively *RY* and *ry*, and consequently the F_1 hybrid zygote formed by their union had the genotype *RrYy*. Now, the crux of the problem lies in the kinds of gametes produced by the F_1 individuals. When the seed surface alone is considered, it is found that the F_1 *Rr* individuals produce gametes half of which carry *R* and half *r*. It is clear, however, that every gamete must necessarily contain within itself not only a gene for seed surface but one for seed colour as well, and, indeed, genes affecting every other character of the plant. Half these same gametes must, therefore, contain the allele *Y* and half the allele *y*; but in any given gamete it is a matter of chance as to whether the gene for round seeds is associated with that for yellow seeds or with that for green seeds. The assortment of the genes is independent. Of that half of the gametes which carry the gene for round seeds, a half in turn (or a quarter of the whole) carry yellow and a half carry green; and of that half that carry the gene for wrinkled seeds, a half also carry yellow and a half green. The F_1 hybrid produces four kinds of gametes in approximately equal numbers: *RY*, *Ry*, *rY*, and *ry*. Now if two such F_1 plants, each of them producing four kinds of gametes, are crossed, there will obviously be 16 possible combinations among their gametes, for there will be four kinds of pollen grains and four kinds of egg cells. The union of these gametes in fertilisation is here, too, a random one, any type of pollen grain being as likely to effect fertilisation as any other; and any type of egg cell being as likely to be fertilized as any other, no selective preference being exhibited between them.

The 16 possible combinations that appear among the F_2 offspring will, therefore, be equally numerous. The parents, the F_1, and the F_2 of the cross that has been used as an example as to both their genotypes and their appearance, the 16 squares in F_2 representing the 16 possible combinations of gametes. A count of these squares makes clear how the 9 : 3 : 3 : 1 ratio arises, for 9 out of these 16 individuals are in appearance round and yellow, 3 are round and green, 3 are wrinkled and yellow, and only 1 is wrinkled and green.

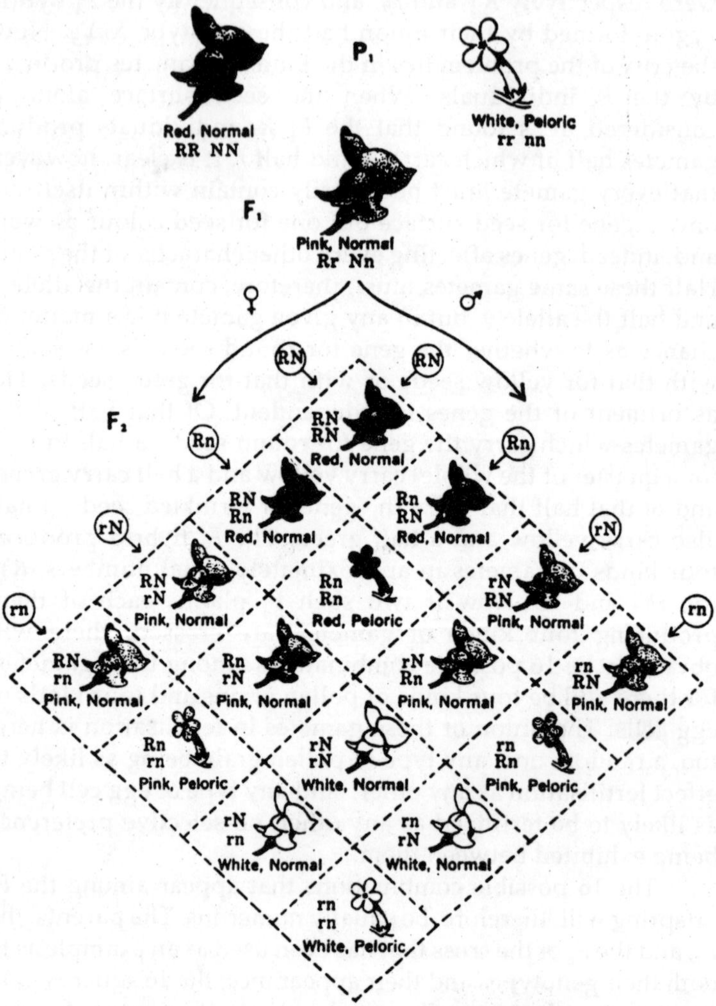

Fig. 9. Diagram showing the independent inheritance in snapdragons of two pairs of characters, in one of which dominance is complete and in theother of which it is lacking. In a cross between a plant homozygous for red flowers of normal shape and one with white and abnormal (pelorice) flowers, the appearance and genotype of the parents, the F_1 and the F_2 are shown.

Back Cross

When F_1 individuals are crossed with one of the two parents from which they were derived, then such a cross is called **back**

cross. In such back crosses, when F_1 is back crossed to the parent with dominant phenotype, no recessive individuals are obtained in the progeny. But when F_1 progeny is back crossed with its recessive parent, both phenotypes (i.e., dominant and recessive) appear in the progeny. While both of these crosses are back crosses, only the cross with the recessive parent is called *test cross.*

Test Cross

A test cross is the mating of an incompletely known genotype to a genotype which is homozygous recessive at all the loci under consideration. The phenotypes of the offspring produced by a test cross reveal the number of different gametes formed by the parental genotype under test. When all of the gametes of an individual are known, the genotype of that individual is also known. A monohybrid test cross gives a 1 : 1 phenotypic ratio, indicating that one pair of factors is segregating. A dihybrid test cross gives a 1:1:1:1 ratio, indicating that two pairs of factors are segregating and assorting independently. A trihybrid test cross, however gives a 1 : 1 : 1 : 1 : 1 : 1 : 1 : 1, ratio, while three pairs of factors are segregating and assorting independently.

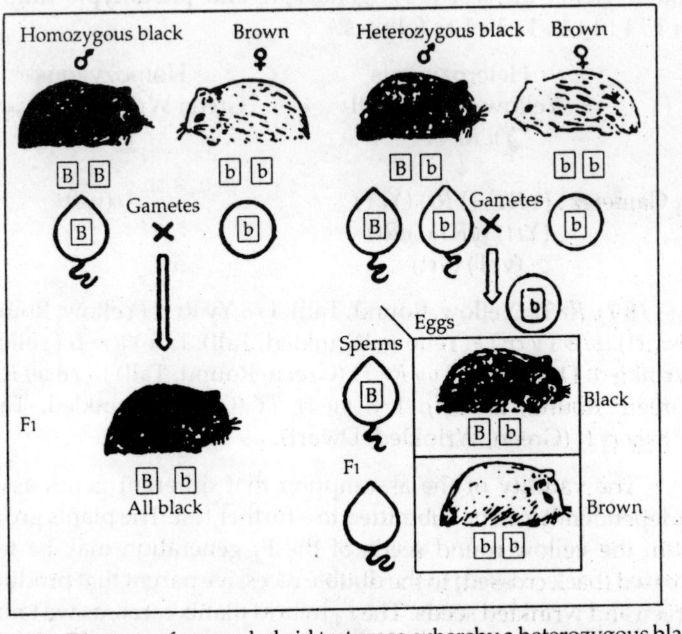

Fig. 10. Diagram of a monohybrid test cross, whereby a heterozygous black guinea pig may be distinguished from a homozygous black one.

Example:

1. ***Monohybrid test cross.*** Test crossing of *Bb* × *bb* produces the 1 : 1 genotypic and phenotypic ratio as has been illustrated in diagram given in Fig. 10.

2. ***Dihybrid test cross.*** The test cross of a heterozygous yellow round seeded pea plant with a double recessive parent (green wrinkled, *yy rr*), yields the test cross genotypic and phenotypic ratio of 1 : 1 : 1 : 1 as follows:

P_1 : Yellow, Round (Heterozygous) *Yy Rr* × Green, Wrinkled (Homozygous recessive) *yy rr*

F_1 : 1 *Yy Rr*: 1 *Yy rr*: 1 *yy Rr* : 1 *yy rr*

or 1/4 Yellow Round : 1/4 Yellow Wrinkled: 1/4 Green Round: 1/4 Green Wrinkled.

3. ***Trihybrid test cross.*** A trihybrid test cross involves crossing of a trihybrid (*Yy Rr Tt*) with its triple recessive (*yy rr tt*) parent and it yields the test cross genotypic and phenotypic ratio of 1 : 1 : 1 : 1 : 1 : 1 : 1 : 1 as follows:

P_1 : Heterozygous Yellow Round Tall *Yy Rr Tt* × Homozygous Green Wrinkled Dwarf *yy rr tt*

↓ ↓

P_1 Gametes : (*YRT*) (*YRt*) (*YrT*) (*Yrt*) (*yRT*) (*yRt*) (yrT) (yrt) — (*yrt*)

F_1: 1/8 *Yy Rr Tt* (Yellow, Round, Tall): 1/8 Yy Rr tt (Yellow, Round, Dwarf): 1/8 *Yy rr Tt* (Yellow, Wrinkled, Tall): 1/8 *Yy rr tt* (Yellow, Wrinkled, Dwarf): 1/8 *yy Rr Tt* (Green, Round, Tall) : 1/8 *yy Rr tt* (Green; Round, Dwarf); 1/8 *yy rr Tt* (Green, wrinkled, Tall): 1/8 *yy rr tt* (Green, Wrinkled, Dwarf).

The validity of the assumption that different genes assort independently can be submitted to a further test. The plants grown from the yellow-round seeds of the F_1 generation may be test-crossed (back crossed) to the double recessive parent that produces green and wrinkled seeds. The F_1 hybrid plants carry, as we know, the genes *Rr Yy* and form four kinds of gametes in equal numbers, namely, gametes with the genes *RY*, *rY*, *Ry* and *ry*. The green-wrinkled parent forms only one kind of gamete, *ry*. The expected result in the progeny of the test cross may, then, be found in the lower most line of squares of the checkerboard. We should obtain yellow-round seeds (genotypically *Rr Yy*), yellow-wrinkled (*rr Yy*), green-round (*Rr yy*), and green-wrinkled ones (*rr yy*) in equal numbers, that is, in a ratio approaching ¼ : ¼ : ¼ : ¼. This is what is actually obtained in experiments, which, accordingly, confirm Mendel's theory. Similarly, one can predict the results to be obtained in Drosophila if the F_1 hybrids of gray-vestigial and ebony long-winged flies are test-crossed to double-recessive ebony-vestigial ones. The F_1 hybrids are phenotypically normal, that is, they have gray body colour and long wings, but they carry the genes for the ebony body colour and for vestigial wings in heterozygous condition. The progeny of the test cross will, then, consist of normal (gray and long-winged), gray vestigial, ebony long-winged, and ebony-vestigial flies in a ratio approximately 1 : 1 : 1 : 1. Test-crossing the F_1 hybrids in summer squashes to plants with yellow spherical fruits will produce segregation in a ratio of 1 white disk: 1 yellow disk: 1 white sphere: 1 yellow spehre. Finally, a test cross of the pink and normal-flowered snapdragons from the F_1 generation of the cross to white peloric plants will produce a segregation in a ratio of 1 pink normal: 1 pink peloric: 1 white normal: 1 white peloric plants.

Trihybrids

When lines differing in three or more independently inheritable characters are crossed, the situation is naturally more complex, but the principle of independent assortment still holds good. Using Mendel's peas, a cross may be arranged between a variety with yellow-round seeds and red flowers and a variety with green-wrinkled seeds and white flowers. Using the same letters for the gene symbols as those used above, the genotypes of the parents may be represented as *RRYYCC* and *rryycc* respectively. Their gametes will contain the genes *RYC* and *ryc*, and the triply heterozygous (trihybrid) F_1 will be *Rr Yy Cc*. Because of dominance, it will have round-yellow seeds and red flowers. In trihybrid eight kinds of gametes will be produced as follows:

RYC	*rYC*
RYc	*rYc*

RyC *ryC*
Ryc *ryc*

In the F_2 generation produced by random union among these eight kinds of gametes, there will evidently be 64 possible combinations, as can be seen by constructing a checkerboard representing the trihybrid cross. Taking dominance into account, it will be found that eight phenotypically different kinds of peas will appear with approximately the following frequencies:

27 round-yellow-red
9 round-yellow-white
9 round-green-red
9 round-green-red
3 round-green-white
3 wrinkled-yellow-white
3 wrinkled-green-red
1 wrinkled-green-white

As in the monohybrid and dihybrid crosses, some of the F_2 individuals that look phenotypically alike will prove to be genotypically diverse. This can be shown by inbreeding them and obtaining an F_3 generation of hybrids.

Polyhybrids

The situations obtained in the offspring of hybrids heterozygous for many genes (*polyhybrids*) may be quite complex. As the number of genes involved in a given cross increases, the number of possible gene combinations increases rapidly. Every added gene multiplies by 2 the number of different classes of gametes formed, by 3 the number of genotypes produced in the F_2 generation, and by the 4 the number of possible combinations of gametes in F_2. As the number of genes involved increases, the chance of recovering one of the original parent types in the F_2 grows rapidly less.

When a single pair of alleles is involved, 1 in 4 of the F_2 will resemble each one of the original parents in appearance and genotype; when two pairs are involved, 1 in 16; when three, 1 in 64; when four, 1 in 256; and so on. When the number of genes (n) involved in a cross reaches tens or hundreds, the total number of genotypes, 3^n, and even the number of homozygous genotypes, 2^n, which may arise in the progeny of a heterozygote become enormous. In fact, these numbers are likely to be greater than the existing number of individuals of a species. It follows that the process of sexual reproduction, with its twin consequences of gene assortment and gene recombination, creates enormous amounts of new genetic variability.

Table 2

The Theoretical Number of Individuals, with their Genotypes and Breeding Behaviour, Expected in F_2 from a Trihybrid Cross of a round, yellow-seeded, red-flowered Variety of pea with a Wrinkled, green-seeded, white-flowered one

Number of Individuals	*Genotype Class*	*Phenotype Class*	*Ratio of Phenotypes*	*Breeding behaviour when self-fertilized*
1	*RR YY CC*			Breeds true
2	*Rr YY CC*			Segregates round-wrinked, 3:1
2	*RR Yy CC*	Round		Segregates yellow-green, 3:1
2	*RR YY Cc*	Yellow	27	Segregates red-white, 3:1
4	*RrYyCC*	Red		Segregates round-wrinkled, yellow-green, 9 : 3 : 3 : 1
4	*Rr YY Cc*			Segregates round-wrinkled, red-white, 9 : 3 : 3 : 1
4	*RR Yy Cc*			Segregates yellow-green, red-white
8	*Rr Yy Cc*			Segregates round-wrinkled, yellow-green, red-white, 27 : 9 : 9 : 9 : 3 : 3 : 3 : 1
1	*RR YY cc*	Round		Breeds true
2	*RR Yy cc*	Yellow	9	Segregates yellow-green, 3 : 1
2	*Rr YY cc*	White		Segregates round-wrinkld 3:1
4	*Rr Yy cc*			Segregates round-wrinkled, yellow-green, 9 : 3 : 3 : 1
1	*RR yy CC*			Breeds true
2	*Rr yy Cc*	Round		Segregates red-white, 3 : 1
2	*RR yy CC*	Green	9	Segregates round-wrinkled, 3 : 1
4	*Rr yy Cc*	Red		Segregates round-wrinkled, red-white, 9 : 3 : 3 : 1
1	*rr YY CC*			Breeds true
2	*rr Yy CC*	Wrinkled		Segregates yellow-green, 3 : 1
2	*rr YY Cc*	Yellow	9	Segregates red-white, 3 : 1
4	*rr Yy Cc*	Red		Segregates yellow-green, red-white, 9 : 3 : 3 : 1
1	*rr yy CC*	Wrinkled		Breeds true
2	*rr yy Cc*	Green	3	Segregates red-white, 3 : 1
		Red		

1	*rr YY cc*	Wrinkled		
2	*rr Yy cc*	Yellow	3	Breeds true
		White		Segregates yellow-green, 3 : 1
1	*RR yy cc*	Round		Breeds true
	Rr yy cc	Greeen	3	Segregates round-wrinkled,
		White		3 : 1
		Wrinkled		
1	*rr yy cc*	Green	1	Breeds true
64		White		

Table 3
The Relation Between the Number of Genes Involved in a Cross and the Number of Phenotypic and Genotypic Classes in F_2

Number of genes involved in the cross	*Number of visibly different F_2 classes of individuals if dominance is complete*	*Number of different kinds of gametes formed by the F_1 hybrid*	*Number of genotypically different combinations*	*Number of possible combinations of F_1 gametes*
1	2	2	3	4
2	4	4	9	16
3	8	8	27	64
4	16	16	81	256
n	2^n	2^n	3^n	4^n

Variations in Mendelian Ratios

One of the most striking things about Mendel's results was the uniform behaviour in inheritance of the seven pairs of contrasting characters used in his crosses. One member of each pair was always dominant and the other always recessive. In a cross of a pure-breeding dominant and a pure-breeding recessive, the F_1 population consisted only of plants with the dominant characteristics. The F_2 of such a cross consisted of both of the original types, in the ratio of three dominants to one recessive. If one of the F_1 plants was crossed with a pure recessive, the ratio of dominants to recessives in the offspring was 1:1. If two pairs of contrasting characters were involved, the F_2 ratio was 9:3:3:1. If the F_1 of the cross just mentioned was crossed to a double recessive, the ratio was 1 : 1 : 1 : 1.

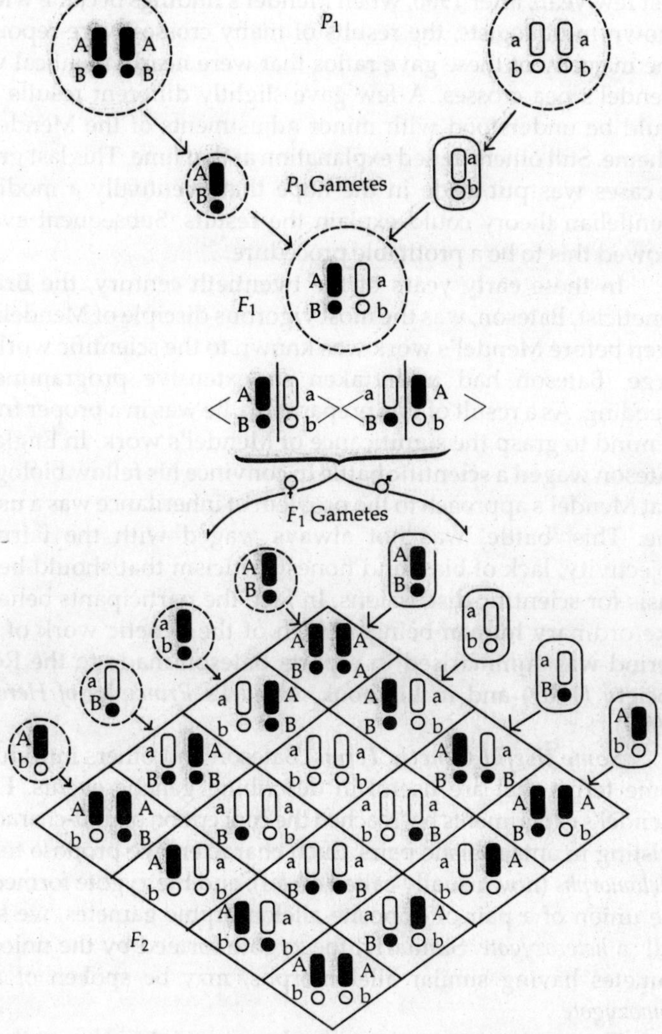

Fig. 11. Diagram showing independent Assortment of two pairs of chromosomes, A-a and B-b. Note that at the meiotic division there are two possible alignments of chromosomes, producing four types of gametes. By random union these produce the sixteen different chromosome combinations shown in the F_2 checkerboard.

Confirmations and Exceptions to Mendel's Scheme. In the first few years after 1900, when Mendel's findings became widely known to biologists, the results of many crosses were reported. The majority of these gave ratios that were nearly identical with Mendel's pea crosses. A few gave slightly different results that could be understood with minor adjustments of the Mendelian scheme. Still others defied explanation at that time. This last group of cases was put aside in the hope that eventually a modified Mendelian theory could explain the results. Subsequent events showed this to be a profitable procedure.

In these early years of the twentieth century, the British geneticist, Bateson, was the most vigorous disciple of Mendelism. Even before Mendel's work was known to the scientific world at large. Bateson had undertaken an extensive programme of breeding. As a result of this preparation, he was in a proper frame of mind to grasp the significance of Mendel's work. In England, Bateson waged a scientific battle to convince his fellow biologists that Mendel's approach to the problem of inheritance was a useful one. This 'battle' was not always waged with the fairness, objectivity, lack of bias, and honest criticism that should be the basis for scientific discussions. In fact, the participants behaved like ordinary human beings. Much of the genetic work of this period was summarised in reports Bateson made to the Royal Society (1909) and in his book, *Mendel's Principles of Heredity* (1909).

Some Useful Genetic Terms. Bateson and others introduced some terms that are useful in describing genetic events. From Mendel's experiments he 'reached the conception of unit-characters existing in antagonistic pairs. Such characters we propose to call *allelomorphs* (now usually called *alleles*), and the zygote formed by the union of a pair of opposite allelomorphic gametes, we shall call a *heterozygote.* Similarly, the zygote formed by the union of gametes having similar allelomorphs, may be spoken of as a *homozygote.*'

To use these terms in examples, we might observe that in peas *round* and *wrinkled* genes are alleles; so are *yellow* and *green* genes. A pure-breeding *round* plant, which we have designated ***RR,*** is homozygous. A *wrinkled* plant, ***rr,*** is also homozygous. The F_1 of a cross of these is ***Rr*** and heterozygous. Recalling that *round* is dominant, we know that a *round* plant can be either homozygous

(*RR*) or heterozygous **(*Rr*)**, and that one could not tell by external appearance which was which. It is worth repeating that there is a difference between appearance and genetic make-up. We speak of the appearance of the organism as its *phenotype* (*yellow* or *green*) and the genetic make-up as its *genotype* (*RR*, *Rr* or *rr*). The individual factors of inheritance, ***R*** or ***r***, carried on the chromosomes are the *genes*.

Returning now to Bateson, these are some of the types of genetic data he gave:

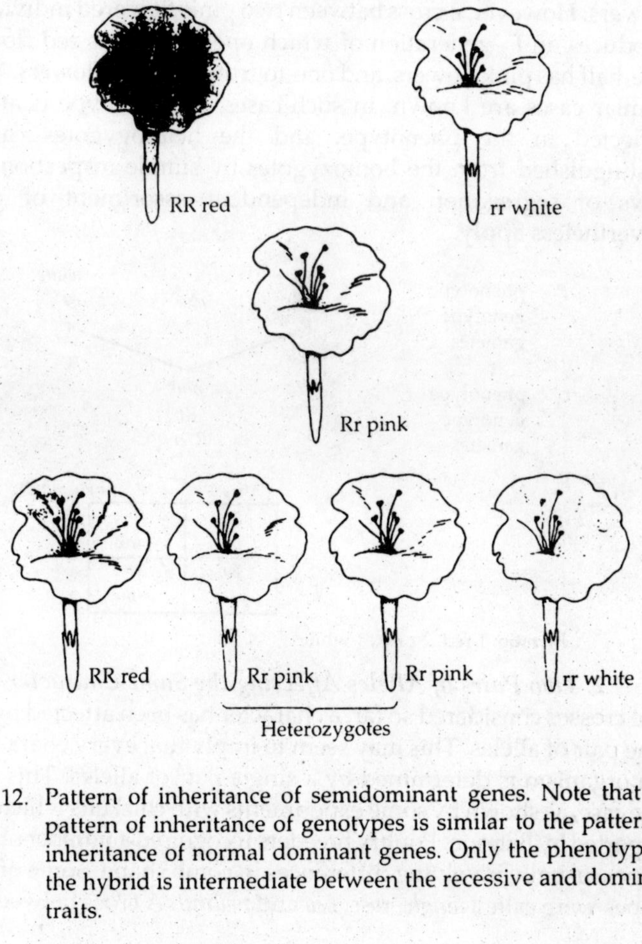

Fig. 12. Pattern of inheritance of semidominant genes. Note that the pattern of inheritance of genotypes is similar to the pattern of inheritance of normal dominant genes. Only the phenotype of the hybrid is intermediate between the recessive and dominant traits.

Exceptions to the Mendelian Ratios

1. Blended Character Expression in Heterozygotes. In Mendel's results the heterozygote was always identical in appearance with the homozygous dominant. Even today werecognise this as the most frequent condition, but there are some cases where the heterozygote is intermediate. In such a situation neither allele is dominant nor recessive. One example is a common cultivated flower, the four o'clock.

In snapdragons, for example, a cross between red- and white-flowered individuals produces offspring all of which have pink flowers. However, a cross between two pink-flowered individuals produces an F_2 generation of which one-fourth has red flowers, one-half has pink flowers, and one-fourth has white flowers. Many similar cases are known. In such cases, the genotype is always reflected in the phenotype, and the heterogygotes can be distinguished from the homozygotes by simple inspection. The laws of segregation and independent assortment of genes nevertheless apply.

P	phenotype	*red*	X	*white*
	genotype	*RR*		*rr*
	gametes	*R*		*r*
F_1	phenotype		*pink*	
	genotype		*Rr*	
	gametes		*R r*	

F_2

	R	*r*
R	*RR* *red*	*Rr* *pink*
r	*Rr* *pink*	*rr* *white*

F_2 ratio: 1 red, 2 pink, 1 white.

2. Two Pairs of Alleles Affecting the Same Character. In all the crosses considered so far, a character has been affected by only one pair of alleles. This may seem to imply that every character of an organism is determined by a single pair of alleles. This is not the case, as shown by some experiments with chickens which were reported by Bateson. Poultry breeders recognised a number of comb types, usually involving differences in comb shape. Some of these types were called *single, rose, pea* and *walnut*. A cross between *rose* and *single* gave *rose* in the F_1 and 3 *rose* to 1 *single* in the F_2. *Rose* and *single* behaved as ordinary alleles with *rose* the dominant.

	phenotype	*rose*	X	*single*
	genotype	*RR*		*rr*
	gametes	*R*		*r*
F_1	phenotype		rose	
	genotype		*Rr*	
	gametes		*R* *r*	

F_2

	R	*r*
R	*RR* *rose*	*Rr* *rose*
r	*Rr* *rose*	*rr* *single*

Similarly, a cross between *pea* and *single* gave *pea* in the F_1 and 3 *pea* to 1 *single* in the F_2. *Pea* and *single* behaved as ordinary alleles with *pea* are dominant.

	phenotype	*pea*	X	*single*
	genotype	*PP*		*pp*
	gametes	*P*		*p*
F_1	phenotype		*pea*	
	genotype		*Pp*	
	gametes		*P* *p*	

F_2

	P	*p*
P	*PP* *pea*	*Pp* *pea*
p	*Pp* *pea*	*pp* *single*

Considered separately there is nothing unusual about these crosses. Considered together there is one puzzle, largely semantic: why is the genotype of *single* written as ***rr*** in the first cross and as ***pp*** in the second? The answer to this question hinges on the fact that the *single* comb condition results from the interaction of two different genes. The two previous crosses and the next one to be described show that both the ***r*** and ***p*** genes are involved in the genetic determination of comb shape.

A comb of the *single* type forms in a chicken when both the *pea* and *rose* genes are recessive. The genotype of a *single* comb chicken, therefore, is ***pprr***. A *pea* comb chicken would have the genotype ***PPrr*** or ***Pprr*** and a *rose* comb chicken would have the genotype ***ppRR*** or ***ppRr***. We might have written the cross of *rose* comb × *single* comb as ***ppRR*** × ***pprr***. At that time, however, we were unaware of the existence of the *p* genes. Furthermore, there was no need to do so since both parents are of the same genotype with respect to the *p* gene: by convention geneticists use the symbols only for those cases where the genes differ in the two parents. If the cross involves differences in both the *p* and *r* comb shape genes, then the necessary genetic symbols must be employed. Such a cross will now be described.

Let us cross the two dominants, *rose* and *pea*. The F_1 will be found to have a new type of comb, *walnut*, which might be looked upon as a blending of the two dominant shapes. The F_2 will give a ratio of 9 *walnut*, 3 *rose*, 3 *pea*, and 1 *single*.

P	phenotype	*pea*	X	*rose*
	genotype	*PP rr*		*pp RR*
	gametes	*Pr*		*pR*
F_1	phenotype		*walnut*	
	genotype		*Pp Rr*	
	gametes		*PR Pr pR pr*	

F_2

	PR	*Pr*	*pR*	*pr*
PR	*PP RR* *walnut*	*PP Rr* *walnut*	*Pp RR* *walnut*	*Pp Rr* *walnut*
Pr	*PP Rr* *walnut*	*PP rr* *pea*	*Pp Rr* *walnut*	*Pp rr* *pea*
pR	*Pp RR* *walnut*	*Pp Rr* *walnut*	*pp RR* *rose*	*pp Rr* *rose*
pr	*Pp Rr* *walnut*	*Pp rr* *pea*	*pp Rr* *rose*	*pp rr* *single*

F_2 ratio: 9 walnut; 3 rose, 3 pea; 1 single

The *single* comb is the result of both (recessive) *p* and *r* genes being homozygous. *Rose* is obtained if the animal is homozygous or heterozygous for *R* and homozygous for *p*. *Pea* is obtained if the animal is homozygous or heterozygous for *P* and homozygous for *r*. *Walnut* results when there is at least one *P* gene together with at least one *R* gene.

3. *Yellow Mice.* Cuénot reported some crosses in mice that neither he nor Bateson could explain in the usual Mendelian manner. He worked with *yellow* and *agouti* hair colour genes. These two types of crosses will demonstrate the problem.

1. *yellow* × *yellow* gives 2 *yellow* to 1 *agouti.*

2. *yellow* × *agouti* gives 1 *yellow* to 1 *agouti.*

Cuénot found it impossible to obtain a strain of *yellow* mice that would breed true. The *yellow* animals always behaved as though they were heterozygous. His *agouti* strains bred true. During the course of his experiments he observed that the litter size in cross of *yellow* × *yellow* was smaller than in the *yellow* × *agouti* cross.

Can you devise a hypothesis to explain the results? From the data given it is possible to arrive at the explanation that was later found to be correct.

4. *Coupling of genes.* Bateson listed another type of result that could not be explained, namely, cases involving two pairs of genes that did not show independent assortment. Let us assume that there are two pairs of genes. **A** and **a** and **B** and **b** and that the phenotypic expression of A is *A*, and that of **a** is *a*, and so on. In the cross **AABB × aabb** we should expect an F_2 phenotypic ratio of 9 *AB*, 3 *Ab*, 3 *aB*, and 1 *ab* if Mendel's rules applied. In some crosses, however, Bateson found that the *A* and *B* characters appeared to be completely *coupled,* that is, they were inherited together. The same was true for *a* and *b*. In fact, the F_2 ratios might be close to 3 *AB* to 1 *ab* with few or no individuals of the *Ab* or *aB* phenotypes. Results of this sort could not be accounted for by the Mendelian scheme but they had been predicted by Sutton.

2

Cell Division

M.J. Schleiden and Theodor Schwann advocated the cell theory in which they stated that all organisms are composed of cell and cell products. Theories on how new cells originate were hotly debated in the decades following Schleiden and Schwann's work. Schleiden and Schwann, among others, thought that the nucleous becomes a new nucleus, which in turn develops into a new cell. Other scientists thought that new cells either beed off from preexisting cells or form by an unknown process in the space between cells. Still others insisted that new cells form when a cell divides into two equal halves. Observations over the years eventually confirmed that cells arise by a process of division, and this idea was well enough established in the 1850s for Rudolph Virchow to complete the formulation of the Cell Theory begun by Schleiden and Schwann. Schleiden and Schwann had said that all organisms are composed of cells and cell products. To this statement, Virchow added that all cells are produced by the division of previously existing cells. The genetic information that determines the characteristics of cells and controls their many functions is contained within chromosomes in cell nuclei.

During cell division, duplicate copies of this genetic material are distributed to the new cells with great precision, thus ensuring the continuity of genetic information from cell to cell. The process by which the genetic material is distributed is called *mitosis.* Mitosis results in two new nuclei, each genetically identical to the original nucleus because each receives a set of chromosomes identical to that of the original cell. The cytoplasm is divided, more or less equally, by a separate process known as *cytokinesis.* In any organism, growth due to an increase in the size of the component cells is limited, and thus extensive growth depends on mitotic cell division. However, cell division does not end when an organism's growth slows or stops at maturity, because cells produced by mitosis also serve as replacements for cells that die or are lost by the organism.

In humans, for example, the outer layers of the skin are constantly being sloughed off and replaced by the division of cells lying beneath them. Continuing cell division that results in cell replacement is vital for the maintenance of homeostasis.

In some types of organisms, specialised cells produced by mitotic cell division may develop directly into new individuals that are genetically identical to the parent. This type of reproduction is called *asexual reproduction.* The other principal method by which organisms reproduce is *sexual reproduction.* Sexual reproduction requires a genetic combination from two different cells and involves genetic contribution from two different cells and involves a type of cell division called *meiosis.*

The Cell Cycle

In the 1870s, the techniques of microscopy had been refined to the point where a number of scientists were able to provide detailed and accurate descriptions of the movement of chromosomes during mitosis. Because there was little visible activity during the remainder of the cell's life span, the processes taking place between mitotic events remained largely unknown. In fact, the period during which mitotic activity was not occurring was viewed as a resting stage, an "interphase". It was not until much later, however, with the development of increasingly sophisticated biochemical techniques in the 1950s, that biologists were able to discover and describe many of the processes that occur in cells between mitotic divisions. They learned that the period between divisions is indeed a very important and active time for the cell. The entire cycle—from the time a new cell is formed until it divides—is called the *cell cycle,* and mitosis can be fully understood only in the context of this cell cycle.

After it had been confirmed that deoxyribonucleic acid (DNA) is the primary genetic material, it was clear that the genetic information contained in DNA must be copied (replicated) before each cell division. Biologists set out to learn exactly when DNA replication takes place. New techniques were developed to study the dynamics of DNA function and duplication throughout the cell cycle. Two major methods of studying DNA are Feulgen staining, which indicates the presence and relative amounts of

DNA, and autoradiography, a process by which DNA can be radioactively labeled and then visualised on photographic film. A combination of these two techniques revealed that the DNA of a cell is replicated during a specific period of the interphase between mitotic divisions, frequently hours before mitosis occurs. The discovery of the timing of DNA replication led to a description of four distinct, functional phases of the cell cycle.

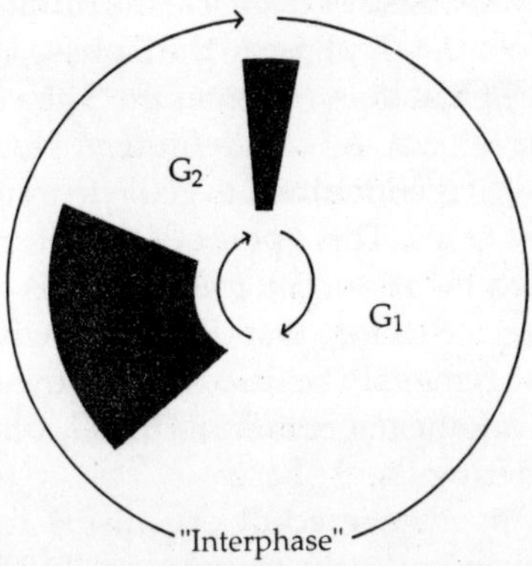

Fig. 1. The cell cycle. Mitosis (*M*) and the synthesis of the genetic material (*S*) are separated by two gap phases (G_1 and G_2).

This phase in which the DNA is synthesized is designated the S (for synthesis) *phase.* The period of time after a cell is produced by mitosis and before it enters the S phase is called the *gap 1 (G_1) phase.* (The G_1 phase is a "gap" only in the sense that it does not include major cell reproductive events involving DNA.) During this G_1 phase, growth and synthesis of many compounds other than DNA occur. After the genetic material is duplicated in the S phase, another gap occurs, the *G_2 phase.* Events in the G_2 phase are not fully understood, but G_2 apparently involves the synthesis of various enzymes and structural proteins necessary for mitosis. The fourth phase, called the *M phase* (for mitosis), is the nuclear division process itself. The length of time a particular cell remains in each stage of the cell cycle varies depending upon the type of cell involved.

In human cells cultured at 37°C, the G_1 stage hours for eight stages, the S phase lasts for six hours, the G_2 phase takes about 4½ hours, and mitosis is completed in just one hour. In the cells of vascular plants, the entire cycle lasts from ten to thirty hours, with the M and S phases comparatively longer than in animals. Many protozoa, fungi and most embryonic cells have much shorter cell cycles with greatly reduced G_1 phases. The length of the G_1 phase is apparently correlated with the amount of growth and synthesis necessary for the cell. Cells that are reproducing very rapidly do not have an extended G_1 period. Furthermore, variations in temperature may increase or decrease the duration of each phase. The transition from the G_1 phase to the S phase is a critical point in the cell cycle. A cell that does not enter the S phase does not divide; in fact, a cell can remain in G_1 indefinitely. However, once a cell enters the S phase, it is committed to go on to mitosis and complete the cycle in most cases. This "point of no return," which occurs late in G_1, is called the *restriction point* and it is a key to the entire cell cycle. The precise factors that determine whether or not a cell enters the S phase remain to be discovered, however. The majority of cells in adult organisms remain in the G_1 phase, permanently blocked from entering the S phase.

Some cells in mature adult organisms continue to divide periodically, or even routinely, to replace cells that are lost or cease to function. Such dividing cells make the critical transition from G_1 to the S phase. A better understanding of what occurs at the restriction point, and thus what determines whether cells enter the S phase and go on to mitosis or remain in G_1, might offer clues for the treatment and control of abnormal cell division. Cancer cells, for example, grow without restraint or control. They are probably derived from cells that are normally blocked from entering the S phase; that is, cells that normally do not divide. Some factor, external or internal, transforms these cells so that they continue to enter the S phase in each cell cycle and thus are committed to uncontrolled cell division. Much of the current basic research on cancer is aimed at determining why cancer cells are not under the constraints that prevent normal cells from dividing in an uncontrolled fashion.

Cell Division in Prokaryotes

Cell division in prokaryotic cells and eukaryotic cells is similar in that in both cases, duplicated genetic material is apportioned to each new cell. In prokaryotes, however, mitosis does not occur. The process of cell division is simplified by the fact that the genetic information is not contained within a nucleus. In bacteria, for example, each copy of the replicated circular chromosome becomes anchored to an attachment point in the cell membrane. The cell membrane elongates until the cell is approximately twice its original length. Then the membrane appears to pinch the cell into two segments, each of which is a new cell that becomes enclosed by a cell wall. Each new cell contains an identical copy of the original bacterial chromosome.

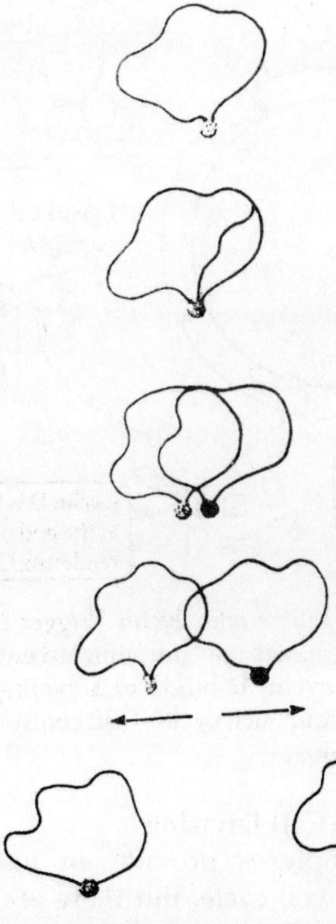

Fig. 2. Cell division in a prokaryotic cell (a) A bacterial cell has a single chromosome which is a circular DNA molecule (with associated RNA and proteins) anchored to the plasma membrane at an attachment point, (b) Replication of the chromosome begins (c) At the completion of replicaton, each chromosome has its own attachment point, (d) Additional plasma membrane is produced, the cell elongates, and the attachment points are moved apart, (e) Two new cells at the completion of cell division.

Mitosis

Mitosis is the division of the nucleus in eukaryotic cells that results in equal distribution of duplicated genetic material to the two new cells (called *daughter cells*). This division is very precise, and the end result of mitosis is two nuclei with genetic information identical to that of the parent cell's nucleus.

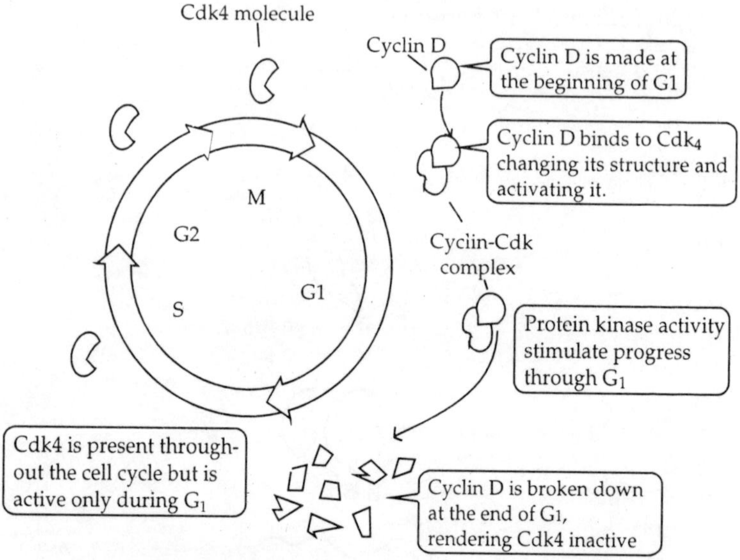

Cyclin-Dependent Kinase and Cyclin Trigger Decisions in the Cell Cycle A huma cell makes the "decision" to enter the cell cell cycle during G1 whne cyclin D binds to a cyclin- dependent kinase (Cdk4). There are four such cyclin-Cdk controls during the typical cell cycle in humans.

(1) Role of Inducer in Cell Division

Cyclin-Cdk complexes provide an *internal* control for progress through the cell cycle. But there are situations in the body where cells that are slowly cycling, or not cycling at all, must be stimulated to divide through *external* controls. When a person is cut and bleeds, specialised cell fragments called platelets gather at the wound and help initiate blood clotting. The platelets also produce and release a protein, *platelet-derived growth factor,* that diffuses to the adjacent cells in the skin and stimulates them to divide and heal the wound.

Other *growth factors* include *interleukins*, which are made by one type of white blood cell and promote cell division in other cells that are essential for the body's immune system defenses, and *erythropoietin*, made by the kidney to stimulate the division of bone marrow cells and the production of red blood cells. In addition, many hormones promote division in specific cells. Growth factors bind to their target cells via specialised receptor proteins on the target cell surface. The specific binding event triggers events within the cell so that it begins a cell cycle. Cancer cells often cycle inappropriately because either they make their own growth factors or they no longer require growth factors to start cycling.

Eukaryotic Chromosomes

Most human cells other than eggs and sperm contain two full sets of genetic information, one from the mother and the other from the father. Eggs and sperm, however, contain only a single set; any particular egg or sperm in your body contains some information from your mother and some from your father. As in prokaryotes, the genetic information consists of molecules of DNA packaged as chromosomes. However, eukaryotes have more than one chromosome, and during interphase these chromosomes reside within the cell's nucleus. Seen through a light microscope, the nucleus appears relatively featureless (except for the dark nucleous) during most of the life of a cell; the chromosomes cannot be seen. Mitosis has been defined historically as the stage.

Although we know less about the organisation of interphase chromatin than about that of mitotic chromatin, we know that interphase chromatin has nucleosomes that are spaced at the same intervals as in supercoiled chromatin. During interphase, DNA thus remains associated with histone molecules while it replicates and directs the synthesis of RNA.

(2) The Stages of Mitosis

Mitosis is a continuous process that is arbitrarily divided into four stages for convenience of description. Four stages of mitosis as it occurs in animal cells are given as follows:

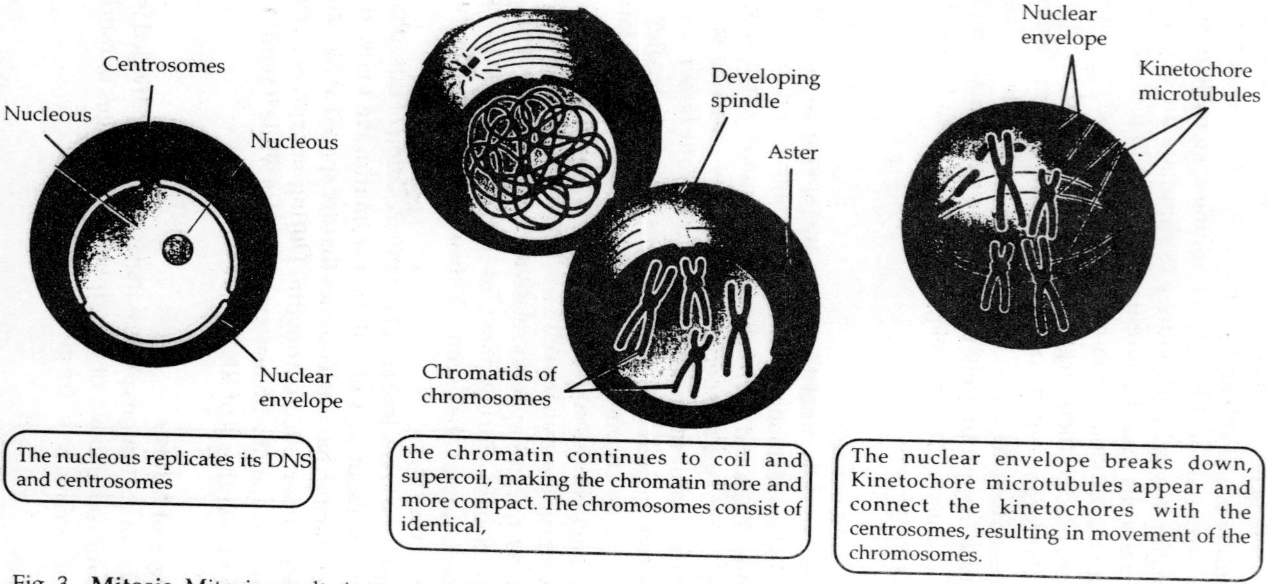

Fig. 3. **Mitosis.** Mitosis results in two new nuclei that are genetically identical to one another and to the nucleus from which they formed. These photomicrographs are of plant nuclei, which lack centrioles and asters. The diagrams are of corresponding phases in animal cells and introduce the structures not found in plants. In the micrographs, the red dye stains microtubules (and thus the spindle); the blue dye stains the chromosomes. In the diagrams, the chromosomes are stylized to emphasize the fates of the individual chromads.

Prophase. The first stage of mitosis is called *prophase.* A typical cell entering prophase from G_2 has an intact nuclear membrane and may have visible nucleoli, but the chromosomes themselves are not visible with a light microscope. During the early stages of prophase, the chromosomes begin to condense. This apparent *condensation* is actually a coiling process that results in a shortening and thickening of each chromosome. The coiling involves not only DNA molecules, but also chromosomal proteins. These protein molecules, particularly *histones,* serve as "spools" around which a thread of DNA is wound.

Chromosome condensation seems to be a necessary part of the mitotic process because chromosomes in their normally extended condition might become quite tangled during mitotic separation. Chromosomes cannot remain permanently condensed, however, because genetic expression (the cell's utilisation of the coded genetic information in the DNA) is complicated when the chromosomes are condensed. Thus, chromosomes coil at the beginning and uncoil at the end of each cell division. In the cells of most organisms, the nuclear membrane begins to break down during prophase. Later, this allows the chromosomes to be apportioned between the two new nuclei that are formed. Also nucleoli in the dividing cell disappear during the early part of prophase. The coiled chromosomes of prophase are easily visible under a light microscope after they have been stained. Each chromosome consists of two subunits that are attached to each other. Each chromosome subunit, called a *chromatid,* contains one doublestranded, helical DNA molecule (as well as various proteins). These two DNA molecules (one in each chromatid) are the products of DNA replication during the S phase of the cell cycle. The point at which the two chromatids are joined is the *centromere,* a distinctly constricted area of the chromosome.

(3) Also during prophase the *spindle apparatus* begins to form. The spindle apparatus consists of a set of microtubules that are associated with chromosome movement later in mitosis. In many kinds of cells, another obvious occurrence at this time is the movement of the *centrioles,* which were duplicated earlier during the S phase of the cell cycle.

During prophase, one pair of centrioles moves to one side of the nucleus and the other pair moves to the opposite side. As they move, the centrioles are accompanied by a halo of microtubules called the *aster*, part of which remains around the centrioles in their new locations on opposite sides of the nucleus.

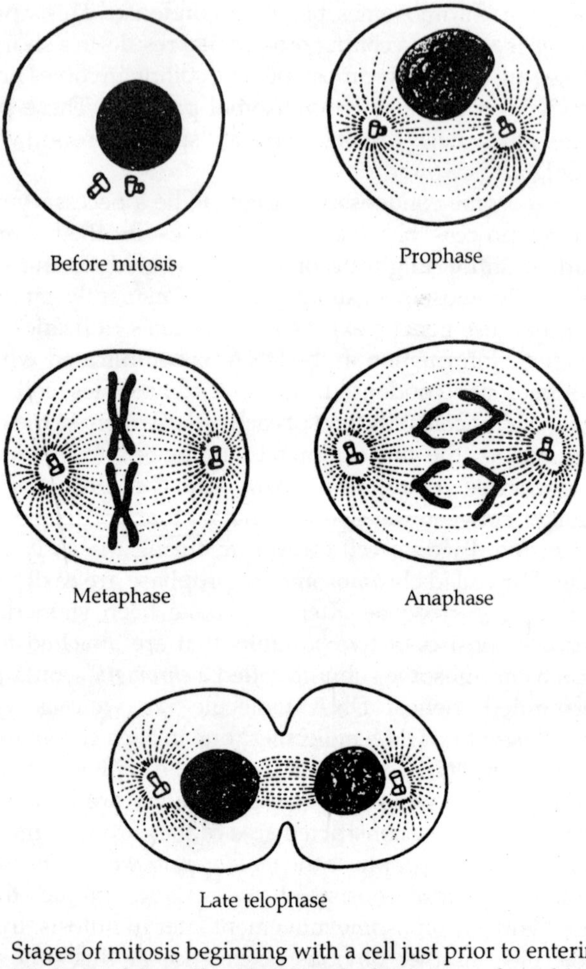

Fig. 4. Stages of mitosis beginning with a cell just prior to entering the M phase of the cell cycle. Note the beginning of cytokinesis in late telophase.

The spindle apparatus develops between these pairs of centrioles. Thus, the centrioles lie near the *poles* (ends) of the spindle apparatus. Despite the proximity of all these structures to one another, there does not seem to be any actual physical connection between the centrioles and the microtubules of either the aster or the spindle apparatus. Because of their prominence and their location in some cells, centrioles were originally thought to be responsible for the organisation of microtubules into the spindle apparatus. However, there is experimental evidence that centrioles do not play a vital role in cell division—removing centrioles from cells that normally have them does not prevent those cells from forming a spindle apparatus and dividing.

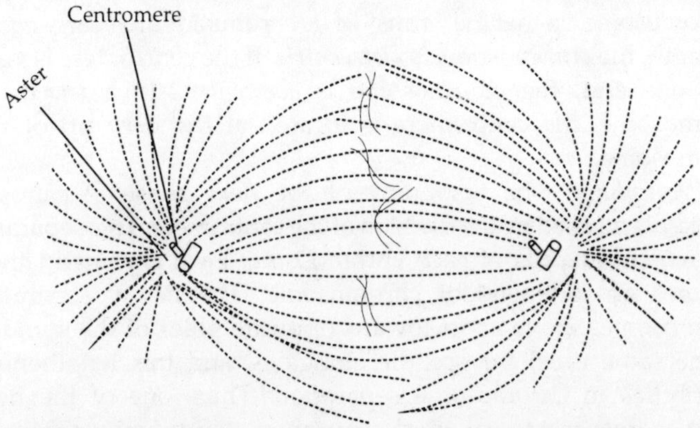

Fig. 5. The spindle apparatus in a dividing animal cell.

Furthermore, cells of many organisms lack centrioles. For example, certain amoebae, the unicellular red algae, some gymnosperms (such as pines), and all angiosperms (flowering plants) lack centrioles altogether. Another feature shared by organisms whose cells lack centrioles is that they do not produce flagellated or ciliated cells. Biologists now think that centrioles are, in fact, identical to basal bodies, the organelles that give rise to cilia and flagella. Thus, it may be that under normal circumstances in organisms whose cells do possess centrioles, the movement of the centrioles during prophase, simply ensures that each new cell will receive a pair of centrioles. During late prophase (sometimes called prometaphase), the chromosomes migrate toward the centre of the cell. By this point the nuclear membrane has completely disappeared.

Metaphase. The next phase of mitosis, called *metaphase,* is a relatively short one. During metaphase, the chromosomes are lined up across the middle of the cell as if on a "plate" that is set perpendicular to the direction of the spindle microtubules. The physical characteristics of chromosomes are most readily observed during metaphase. For this reason, metaphase chromosomes are used to study individual chromosomes within single cells and to make comparisons between chromosomes of different species.

During metaphase it is especially easy to see that chromosomes differ not only in size, but also in the relative position of their centromeres. If the centromere is located near the centre of a chromosome, so that the "arms" of its chromatids are nearly equal in length, the chromosome is *metacentric.* If the centromere is very near one end, the chromosome is *acrocentric.* In a *telocentric* chromosome, the centromere is located at the very tip of the chromosome.

Anaphase. The beginning of the next phase of mitosis, *anaphase* is marked by the division of the centromere. This separates the two chromatids of each chromosome. Each chromatid then becomes an independent chromosome. The newly separated chromosomes move apart toward opposite poles of the spindle. At the same time, the spindle elongates, and this lengthening contributes to chromosome separation. Thus, one of the new chromosomes produced when centromere divides moves toward one end of the cell, and the other moves toward the opposite end. Neither the initial separation of the chromosomes during anaphase nor their subsequent movement toward the poles of the spindle is well understood, but the microtubules of the spindle almost certainly are involved in this movement.

Microtubules also seem to be responsible for the overall lengthening of the spindle. There are two sets of microtubules in the spindle. Some microtubules attach to *kinetochores,* multilayered structures that develop on each of two opposite faces of the centromere of each chromosome. The microtubules attached to the kinetochore on one side of the centromere stretch from the centromere to one pole (end) of the spindle apparatus. Those attached to the kinetochore on the other side of the centromere stretch to the opposite pole. These microtubules are called *centromere-to-pole* microtubules. A second set of longer microtubules, not attached to the chromosomes, stretch further and overlap with one another to produce what appear to be *pole-to-pole* connections. It is generally thought that a major force involved in chromosome movement is produced by actions of the centromere-to-pole microtubules that are attached to the centromere of each of the newly separated chromosomes. As a chromosome moves, its "arms" lag slightly behind, as if the chromosome were being pulled by the microtubules attached to it. This apparent pulling results in certain characteristic chromosome shapes. For instance, metacentric chromosomes assume a V shape and acrocentric chromosomes a J shape as they move through the cytoplasm, with the apex of the V or J being the centromere, where spindle microtubules are attached.

The mechanism of chromosome movement is still the subject of much investigation and some debate. One mechanism proposed for chromosome movement is the tubule assembly-disassembly hypothesis, which suggests that the centromere-to-pole microtubules are disassembled into their constituent subunits at the poles of the spindle apparatus. This would shorten the microtubules and pull the attached chromosomes toward each pole. The released subunits might then be added to the pole-to-pole microtubules that are not attached to chromosomes but instead overlap to stretch from one end of the spindle to the other. Together, these processes would pull the chromosomes toward the poles and, at the same time, add to the total length of the spindle. Another hypothesis regarding chromosome movement is the sliding microtubule mechanism. This hypothesis suggests that the chromosomes move as the chromosome-to-pole and pole-to-pole microtubules slide past each other.

This sliding process would result from chemical bonds forming, breaking, and reforming between spindle fibers that are moving in opposite directions—in a manner similar to the sliding of fibres in contracting muscle cells. Such microtubule sliding might also explain the overall lengthening of the spindle during anaphase. Some biologists propose yet a third possibility for chromosome movement. Because of the discovery that actin and myosin are present in the spindle, these biologists hypothesize that the microtubules provide only the structural framework of the spindle. They suggest that movements of microtubules (and chromosomes) are due to interactions of actin and myosin, as are several other types of cell movement. There is some evidence opposing this hypothesis, however, because inhibitors that prevent actin molecules from binding with other molecules do not prevent chromosome movements during anaphase.

Telophase. It is characterised by the formation of nuclear membrane and cytokinesis. When the chromosomes arrive at the poles, the final phase of mitosis, *telophase,* begins. During this phase the spindle apparatus disintegrates, the chromosomes begin to uncoil, and the two new nuclei organise. This nuclear reorganisation involves the assembly of nuclear membranes and the reappearance of the nucleoli.

1. Distributing exact Copies of Genetic Information

Eukaryotic cells contain multiple chromosomes. Mitosis ensures the accurate distribution of these chromosomes to the daughter nuclei. During mitosis, a single nucleus gives rise to two nuclei that are genetically identical to each other and to the parent nucleus. In reality, mitosis is a continuous process in which each event flows smoothly into the next. For discussion, however, it is convenient to look at mitosis—the M phase of the cell cycle—as a series of separate events, or subphase, prophase, prometaphase, metaphase, anaphase and telophase.

During interphase, the nucleus is between divisions. At this stage we see the nuclear envelope, the nucleoli, and a barely discernible tangle of chromatin. Once the decision to enter mitosis has been made, DNA replicates, and during G2 a pair of *centrosomes* forms from a single centrosome that lies near the nucleus. The centrosomes are regions of the cell that initiate the formation of microtubules and thus help orchestrate chromosomal movement. The regions are not enclosed by membranes and are not visible as discrete objects, but their positions are made evident by the arrangement of nearby microtubules. In many organisms, each centrosome contains a pair of *centrioles.* However, the centrosomes of seed plants and some other organisms do not have centrioles. Where present, each centriole pair consists of one "parent" and one smaller "daughter" centriole at right angles to the parent centriole.

2. Formation of Spindle

The appearance of the nucleus changes as the cell enters *prophase*—the beginning of mitosis. The centrosomes move away from each other toward opposite ends, or *poles,* of the cell. Each centrosome then serves as a *mitotic centre* that organizes microtubules. Some of the microtubules, called *polar microtubules,* run between the mitotic centres and make up the developing *spindle.*

The spindle is actually two *half spindles*: Each polar microtubule runs from one mitotic centre to the middle of the spindle, where it overlaps with polar microtubules of the other half spindle. Polar microtubules are initially unstable, constantly forming and falling apart until they contact polar microtubules from the other half spindle and become more stable. The spindle is responsible for the movement and distribution of chromosomes during mitosis and meiosis.

3. A Prophase Chromosome

The chromatin also changes during prophase. The extremely long, thin fibers take on a more orderly form as a result of coiling and compacting. Under the light microscope, each prophase chromosome is seen to consist of two chromatids held tightly together over much of their length. The two chromatids of a single mitotic chromosome are identical in structure, chemistry, and the hereditary information they carry because of the way in which DNA replicates during the S phase that precedes mitosis.

Within the region of tight binding of the chromatids lies the centromere, which is where chromatids become associated with the microtubules of the spindle. Very late in prophase, specialised three-layered structures called *kinetochores* develop in the centromere region, one on each chromatid. This is where the microtubules actually attach.

4. Movements of Chromosomes

The somewhat condensed chromosomes start to move at the end of prophase, which is the beginning of *prometaphase.* The nuclear lamina disintegrates and the nuclear envelope breaks into small vesicles, allowing the developing spindle to "invade" the nuclear region. Some polar microtubules now attach to chromosomes at their kinetochores and so are called *kinetochore microtubules.* The kinetochore of one chromatid is atttached to microtubules coming from one pole, while the kinetochore of its sister chromatid is attached to microtubules emanating from the other pole.

At the attachment points are proteins that act as "molecular motors"; these *motor proteins* have the ability to hydrolyze ATP to ADP and phosphate, thus releasing energy to move the

Equatorial (metaphase) plate
The duplicated centromere regions connecting paired chromatids become aligned in a plane at the cell's equator.
Daughter chromosomes
The duplicated centromere regions connecting paired chromatids become aligned in a plane at the cell's equator.
The separating chromosomes reach the poles. Telophase passes into the next interphase as the nuclear envelopes and nucleoli reform and the chromatin becomes diffuse.

Fig. 6.

chromosomes along microtubule "railroad tracks" toward the poles. The movement of chromosomes toward the poles is counteracted by two factors. First, there seems to be a repulsive force that keeps the chromosomes in the middle of the spindle region. Second, the two chromatids are held together at their centromere.

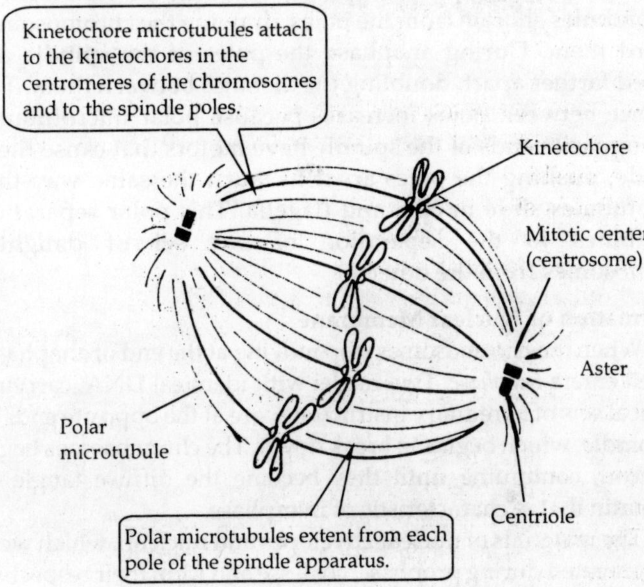

Fig. 7. The Mitotic Spindle Consists of Microtubules
Diagram of the spindle apparatus in a cell at metaphase.

Chromosomes during prometaphase appear to move aimlessly back and forth between the centrosomes at the poles and the middle of the spindle. Gradually, the kinetochores approach this middle region, called the *equatorial plate* or *metaphase plate,* halfway between the poles. The cell is said to be in *metaphase* when all the Kinetochores arrive at the equatorial plate. Metaphase lasts up to an hour and is the best time to see the sizes and shapes of chromosomes. At the end of metaphase the centromeres separate, possibly because a protein holding the chromatids together breaks down. Separation of the centromeres marks the beginning of *anaphase,* the phase of mitosis during which the two sister chromatids of each chromosome—now called *daughter chromosomes,* each containing one double-stranded DNA molecule—move to opposite ends of the spindle. What propels this highly organised mass migration, which takes about 10 minutes, is not clear.

Two things seem to move the chromosomes along. First, motor proteins propel them, and second, the kinetochore microtubules shorten from the poles, drawing the chromosomes toward them. During anaphase the poles of the spindle are pushed farther apart, doubling the distance between them. The distance between poles increases because polar microtubules from opposite ends of the spindle have motors that cause them to slide, pushing the poles apart in much the same way that microtubules slide in cilia and flagella. This polar separation contributes to the separation of one set of daughter chromosomes from the other.

5. Formation of Nuclear Membrane

When the chromosomes stop moving at the end of anaphase, the cell enters *telophase.* Two nuclei with identical DNA, carrying identical sets of hereditary instructions, are at the opposite ends of the spindle, which begins to break down. The chromosomes begin to uncoil, continuing until they become the diffuse tangle of chromatin that is characteristic of interphase.

The materials of nuclear envelopes and nucleoli, which were disaggregated during prophase, coalesce and form their respective structures. When these and other changes are complete, telophase—and mitosis—is at an end, and each of the daughter nuclei enters another interphase.

During interphase, the DNA duplicates and new chromatids form, so that each chromosome consists of two chromatids. Centrioles, if present, replicate during interphase: The two paired centrioles, first separate, and then each acts as a "parent" for the formation of a new "daughter" centriole at right angles to it. Mitosis is beautifully precise. Its result is two nuclei that are *identical to each other* and to the parent nucleus in chromosomal makeup and hence in genetic constitution.

Cytokinesis

Mitosis refers only to the division of the nucleus. The division of the cell's cytoplasm is accomplished by *cytokinesis.* Cytokinesis generally, but not always, follows mitosis. Animal cells usually divide by a furrowing of the membrane, as if an invisible thread were tightening between the two parts. The "invisible" threads are microfilaments of actin and myosin located in a ring just beneath the plasma membrane. These two proteins interact to produce a contraction, just as they do in muscles, thus pinching the cell in two.

In plant cell the cytoplasm divides differently, because plants have cell walls. As the spindle breaks down after mitosis, membranous vesicles derived from the Golgi apparatus appear in the equatorial region roughly midway between the two daughter nuclei of a dividing plant cell. With the help of microtubules, the vesicles fuse to form new plasma membrane and contribute their contents to a cell plate, which is the beginning of a new cell wall. Following cytokinesis, both daughter cells contain all the components of a complete cell.

The precise distribution of chromosomes is ensured by mitosis. Organelles such as ribosomes, mitochondria, and chloroplasts need not be distributed equally between daughter cells, as long as some of each are present in both cells; accordingly, there is no mechanism with precision comparable to that of mitosis to provide for their equal allocation to daughter cells.

Number of Chromosomes

The nuclei of most cells in the body of any given animal or vascular plant contain the same number of chromosomes, a number that is the same in the cells of all members of that particular species. This number is the *diploid* number of chromosomes and it is symbolized 2N. Each species has its own characteristic diploid chromosome number. For example, alligators have 32 chromosomes, humans have 46, dogs 78, tomatoes 12, and soybeans 40. These diploid chromosome sets are made up of pairs of like chromosomes. Human cells contain 23 pairs of chromosomes, tomato cells 6 pairs, and soybeans 20 pairs. The two members of each of these chromosome pairs are known as *homologous chromosomes*. Homologous chromosomes are identical to each other in appearance, the two members of a homologous chromosome pair both bear genetic information that controls the same characteristics.

In mitosis, homologous chromosomes act independently of each other. Chromosomes are randomly arranged on the metaphase plate, and a given chromosome is no more likely to be situated near the other member of the homologous pair to which it belongs than it is to be near any of the other chromosomes. We emphasize this characteristic of chromosome behaviour in mitosis at this point because it contrasts sharply with the situation we will encounter when we examine chromosome behaviour in meiosis.

As stated already that the specialised reproductive cells such as eggs and sperm have exactly half the diploid number of chromosomes. The chromosome number characteristic of such cells is called the *haploid* number, and it is symbolized 'N'. Thus, the eggs and sperm of alligators contain 16 chromosomes, those of dogs 39, and so forth. Such haploid cells contain only one member of each pair of homologous chromosomes.

Meiosis

Sexual reproduction requires a genetic contribution from two different cells. These two specialised reproductive cells, called *gametes,* combine in a process known as *fertilisation.* This union of gametes forms a single cell, the *zygote,* which is the first cell of a new organism. Each of the two gametes contributes half of the genetic information for the new organism. You can see, then, that before fertilisation, organisms that reproduce sexually *must* produce gametes with half the normal chromosome number of their ordinary body cells. If this did not occur, the fusion of the two gametes would result in a zygote with twice the parents' number of chromosomes, and the number of chromosomes would continue to double with each generation.

Reduction in chromosome number is accomplished by a type of cell division called *meiosis.* Thus, these two processes—the doubling of the chromosome number by fertilisation and the halving of the chromosome number by meiosis—ensure that the number of chromosomes in a given kind of organism is stable from generation to generation.

Meiosis reduces the number of chromosomes to one-half of the diploid number. As we noted earlier, cells that contain this reduced number of chromosomes are said to be haploid (N), or to have a haploid number of chromosomes. Therefore, in an organism with a diploid chromosome number of eight, the haploid number of chromosomes contained in the gametes is four. Human body cells have forty-six chromosomes, while human eggs and sperm,

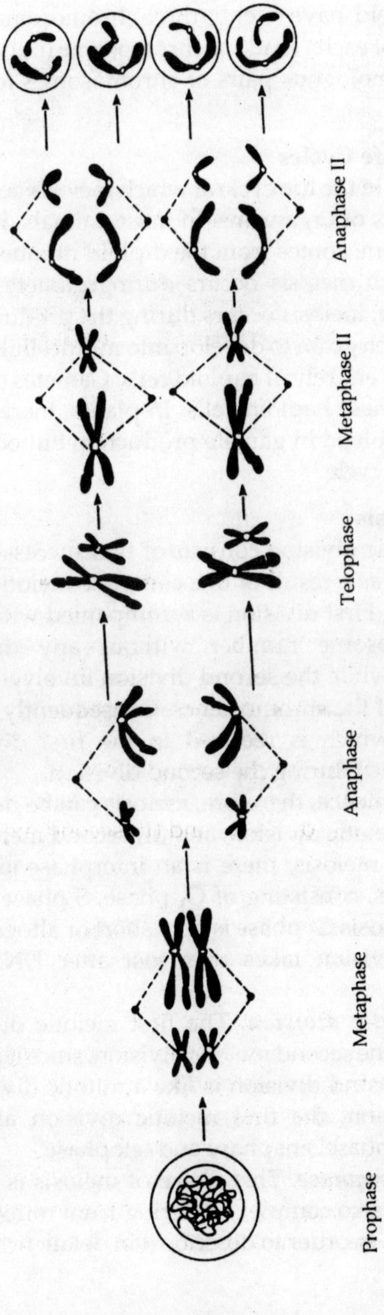

Fig. 8. Stages of meiosis.

which are haploid, have twenty-three chromosomes. Haploid cells have only one of each "kind" of chromosome (only one member of each of the homologous pairs of chromosomes found in diploid cells).

Meiosis and Life Cycles

The point in the life cycle at which meiosis occurs is different in various types of organisms. In most animals, reduction in the number of chromosomes from the diploid number to the haploid number through meiosis occurs during gamete production. In plants, however, meiosis occurs during the production of haploid (N) spores, which go on to develop into multicellular reproductive bodies made up entirely of haploid cells. Gametes develop directly from some of these haploid cells. In plants, therefore, meiosis is not directly involved in gamete production but occurs elsewhere in the plant life cycle.

Stages of Meiosis

The meiotic division consists of two successive divisions of the cell, so that as a result of one complete meiotic division, four cells will result. First division is accompanied with the reduction in the chromosome number without any division of the chromosomes, while the second division involves the separation of chromatids of the chromosomes. Consequently, the number of chromosomes which is reduced in the first division remains constant (haploid) during the second division.

For convenience, therefore, meiosis can be described in two parts: (i) first meiotic division, and (ii) second meiotic division.

Preceding meiosis, there is an interphase just like the one found in mitosis, consisting of G_1 phase, S phase and G_2 phase. However in meiosis G_2 phase is very short or altogether absent, so that meiotic division takes over just after DNA synthesis is complete.

First meiotic division. The first meiotic division is more important than the second meiotic division, since it is the reduction division. The second division is like a mitotic division. Like any other cell division, the first meiotic division also consists of prophase, metaphase, anaphase and telophase.

(a) First prophase. This phase of meiosis is of a very long duration and is also complex, differing from mitotic prophase in several respects. In order to describe it in detail, first prophase has been subdivided into five stages, namely, leptotene, zygotene, pachytene, diplotene and diakinesis. These five stages are established for the convenience of description. Otherwise such a distinction is not possible and one stage very gradually passes on into the next stage.

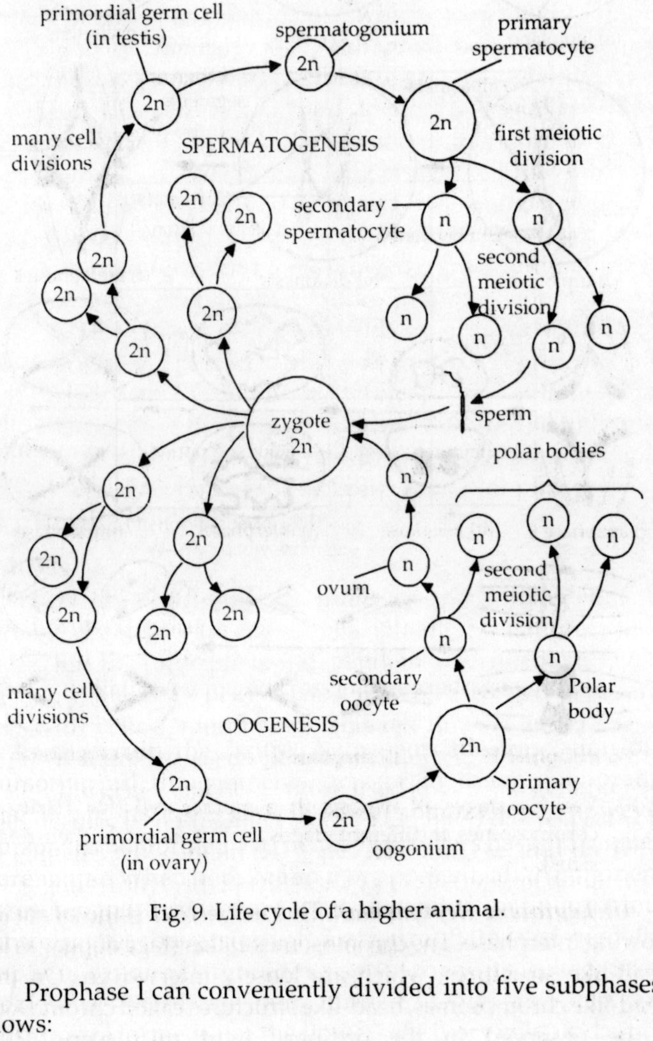

Fig. 9. Life cycle of a higher animal.

Prophase I can conveniently divided into five subphases as follows:

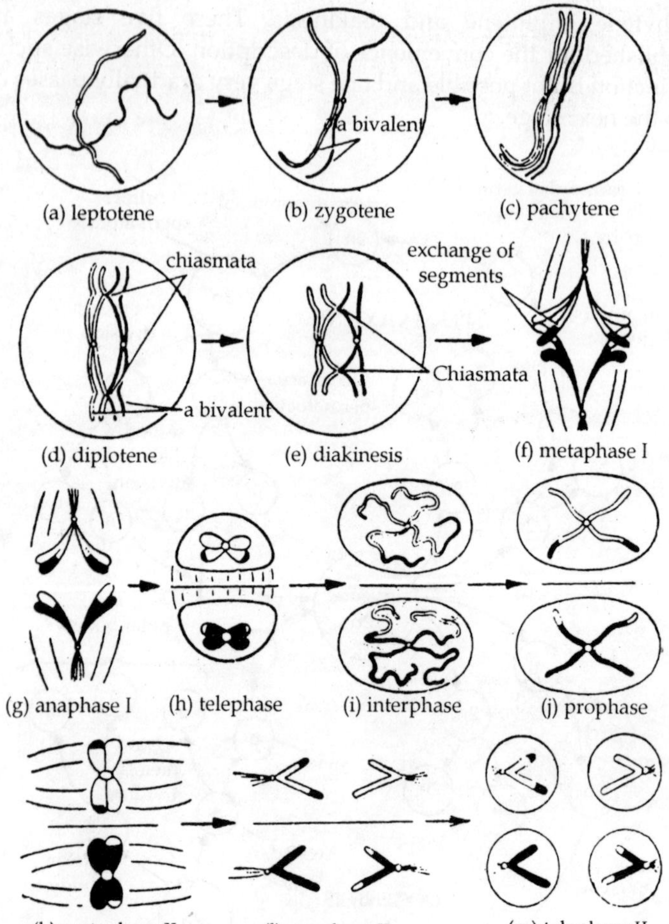

Fig. 10. Various stages of meiosis in a mother cell (for clarity, the chromosomes at different stages are not shown in their relative sizes).

(i) Leptotene (leptonema). This is the first stage of meiosis following interphase. The chromosomes at this stage appear as long thread-like structures, which are loosely interwoven. On these thread-like chromosomes, bead-like structures called chromosomes can be observed in the ordinary light microscope. These chromomeres are found all along the length of the chromosomes. In some animals, particularly in insects, the chromosomes have their ends drawn together at a point on the nuclear membrane, near the centriole. Absence of such polarisation of chromosome ends in plants is perhaps due to the absence of centrosome.

In certain plants like *Lilium* the chromosomes are densely clumped to one side, no chromatin material being seen in the rest of the nucleus. This phenomenon is called 'synizesis' and its significance is not clearly understood. Although it is believed that the duplication of chromosomes takes place during leptotene, but under light microscope there is hardly any evidence of double nature of the chromosomes. However, as we know, DNA synthesis actually takes place during the S-shape of premeiotic interphase.

(ii) Zygotene (zygonema). This sub-phase is characterised by the pairing of homologous chromosomes (synapsis). The pairing is brought about in a zipper-like fashion and may start at the centromere, at the chromosome ends or at any other position. This pairing takes place between homologous segments even if they are present in the non-homologous chromosomes. Another important feature of this pairing is that pairing is allowed only between two chromosomes in one region.

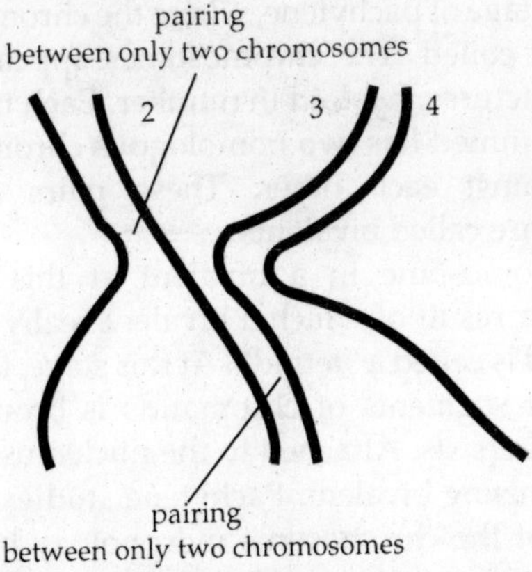

Fig. 11. Meiotic pairing among four homologous chromosomes.

In an autotetraploid, where there will be four homologous chromosomes, in a particular region only two chromosomes will pair. It seems that as soon as two chromosomes, come in contact, their attraction forces are neutralised. The only exception is the pairing between the giant chromosomes in the somatic cells of the salivary glands of *Drosophila.* In this case, if there are three or more homologous chromosomes, these will pair throughout their length, but this is not meiotic pairing. An important question at this stage, is what are those factors which bring about pairing of chromosomes in meiotic prophase? This aspect is being currently studied at the biochemical level.

Darlington (1930) gave an explanation for meiotic chromosome pairing in the form of "precocity theory". According to this theory, meiosis is a precocious mitosis as a result of which the chromosomes which have not yet duplicated, have to enter prophase. This is responsible for chromosome pairing. However, recent work has shown that DNA synthesis is completed before the onset of meiosis. Therefore, the precocity theory is no longer tenable. Recently Stern and Hotta (1969) have shown that the premeiotic S-phase, unlike premitotic one, does not include total chromosome replication. About 0.3 per cent DNA, forming a significant component, replicates only at about the zygotene and is believed to control chromosome pairing.

(iii) Pachytene (pachynema). Once the homologous chromosomes have undergone pairing at the zygotene stage, the cell enters the stage of pachytene, where the chromosomes become shortened and coiled. The chromosomes appears as thickened thread-like structures, haploid in number. Each thread, however, if carefully examined has two homologous chromosomes closely appressed against each other. These pairs of homologous chromosomes are called bivalents.

Each chromosome in a bivalent at this stage has two chromatids as a result of which a bivalent really consists of four chromatids and is called a "tetrad". At this stage, the crossing over or exchange of segments of chromatids is brought about. The nucleolus still persists. Attached to the nucleolus can be seen the nucleolar organising bivalent. Pachytene studies are very useful for the study of the chromosome morphology because they are longer than the chromosomes in mitotic metaphase, so that the structural details of chromosomes can be easily resolved.

(iv) Diplotene (diplonema). At diplotene, further thickening and shortening of the chromosomes take place. The homologous chromosomes start separating from one another. This separation starts at the centrosomes, and travels along towards the ends. This kind of separation from centromere towards the ends is known as "terminalisation". Due to such separation, the dual nature of a bivalent becomes distinct and hence the name diplotene. The homologous chromosomes are now held together only at certain points along the length. Such points of contact between the homologous chromosomes are known as chiasmata (singular, chiasma) and represent the places of crossing over (actual exchange of chromosome segments). As the terminalisation occurs, these chiasmata move towards the ends of the chromosomes.

(a) (b) (c)

Fig. 12. Terminalisation of chiasmata during prophase I.

During crossing over at one point, only two chromatids, one from each of the two homologous chromosomes, take part. It should be emphasized that the chiasmata are not the cause, but are only the consequence of crossing over. There are various views explaining the mechanism of crossing over and chiasma formation.

The number of chiasmata per bivalent is normally dependent on the length of the chromosomes. These can be terminal when found at the end of bivalent due to terminalisation effect or can be intercalary. The terminal chiasmata also originate at intercalary positions. At the diplotene stage, the chromosomes undergo shortening. The nucleolus is also diminished but still persists. No nuclear membrane can be seen at this stage.

(v) Diakinesis. It is the last prophase of meiosis. At diakinesis, the chromosomes continue to undergo further contraction. Ordinarily the only distinction between diplotene and diakinesis is the more contracted state of bivalents at diakinesis. The nucleolus may not be seen at this stage. Due to further terminalisation and contraction, the bivalents appear as rounded bodies darkly stained (with acetocarmine) and evenly scattered throughout the cell. Consequently; the chiasmata are mainly terminal and chromosome counting is easy at this stage. In case of longer chromosomes with intercalary chiasmata, all chiasmata are not terminalised, some of them still being in the intercalary region.

The bivalents at this stage have a tendency to be at the periphery near the nuclear envelope. It is only during the later part of this stage or during the early metaphase I, that the nuclear membrane disappears.

Metaphase I

It is similar to the metaphase of mitosis in that the chromosomes are lined up in the centre of the cell. The orientation of each chromosome, however, is different from that seen in mitosis because of the meiotic pairing of homologous chromosomes. In mitotic metaphase, the centromeres of the chromosomes are lined up along the equatorial plane, while the arms of the chromosomes point in various directions. In metaphase of the first meiotic division, however, chromosomes are specifically oriented so that centromere regions of homologous chromosomes point toward opposite poles of the spindle apparatus.

Anaphase I

It begins when homologous chromosomes separate and begin to move toward each pole. Because the chromosomes of each homologous pair separate and move in opposite directions, each daughter cell receives one chromosome from each homologous pair. However, chromatids do not part at this stage because there is no centromere division in the first meiotic division. Thus, each of these chromosomes still consists of two chromatids and is called a *dyad.* Following the first meiotic division, then, the number of chromosomes has been reduced to half of the diploid number; the resulting nuclei are now haploid. However, because each chromosome still consists of two chromatids (recall that there was no centromere division in anaphase I), one more division is required to separate the chromatids. It is during the second meiotic division that centromeres divide and chromatids separate.

Telophase I

In some organisms there is a *telophase I* stage at the end of the first division. In this stage, as in telophase of mitosis, the nuclear envelope reforms and the nucleoli reappear. The chromosomes may also uncoil. In other organisms, however, there is no telophase at the end of the first meiotic division, and chromosomes which are grouped at anaphase I enter directly into the second division.

In either case, there is no DNA replication (S phase) between the two meiotic divisions.

The Second Meiotic Division

The phases in the second division of meiosis, are referred to as *prophase II, metaphase II, anaphase II,* and *telophase II.* Centromere division finally occurs during anaphase II, and the chromatids part to become independent chromosomes. At the end of telophase II, the final products of the two divisions of meiosis are four new cells. Each of these four cells is haploid; that is, it has half the diploid number of chromosomes. Because of the crossing over that has occurred, together with the assorting of homologous chromosomes into different associations, the meiotic products are genetically different, not only from the original parent cell, but also from each other.

Synaptonemal Complex (SC) and its Significance in Meiosis

Moses in 1956 first discovered Synaptonemal Complex (SC), a feature of meiotic prophase. Synaptonemal Complex is a tripartite structure usually found between the two paired homologous chromosomes of each bivalent in all animal and plant nuclei undergoing meiosis. This is considered to be a physical structure which is associated with synapsis of homologous chromosomes. Complete synaptonemal complexes are seen at zygonema in the region of pairing. At pachynema these complexes are even more conspicuous. The synaptonemal complex is a configuration in which the 10 nm fibres of the chromosomes are arranged into a superstructure that can be viewed under the electron microscope. It is composed of three parallel, electron dense elements that are separated by less dense areas. The two lateral elements seem to be composed of fibres that are slightly wider than 10 nm and are called *synaptomeres.* They vary in structure between different stages of meiotic prophase I within a species. The central element is a ladder like configuration in the centre of the SC.

In some species it is comparatively more pronounced. The transverse elements have electron-dense filaments that interconnect the central element with the lateral elements. The lateral elements may be spaced from 20 nm to 30 nm to as much as 100 nm to 125 nm. Cytochemical studies have demonstrated that the lateral elements are rich in DNA, RNA and proteins, but that the central elements contains mainly RNA, protein and little DNA.

Sometimes, some synaptonemal complex like structures are also found which are not associated with synapsed chromosomes. They show some similarities with the true synaptonemal complex. The lateral elements of these anomalous complexes may differ in size and density from the true synaptonemal complexes of that species.

The true autosomal synaptonemal complex, is always unit tripartite complex but anomalous complexes are often multiple stalks of alternating lateral elements and central elements. In 1970, King presented a hypothesis for the formation of synaptonemal complex. The hypothesis is popularly known as *"The Synaptomere—Zygosome Hypothesis."* According to this hypothesis, there are structures called synaptomeres which are coiled polynucleotide segments scattered along the length of a pair of synapsed chromosomes lying in close proximity to one another. Each synaptomere is composed of three segments: A, B and C. The lateral segments of the synaptomeres (A and C) pair with the respective segments of the adjacent synaptomeres. The B segments are directed towards the central element and are the sites where the so-called zygosomes are attached. Zygosomes are rod shaped subunits, assembled in the nucleoplasm and are each visualised as protein molecules having a folded head end by which they can attach to the central segment (B) of the synoptomere.

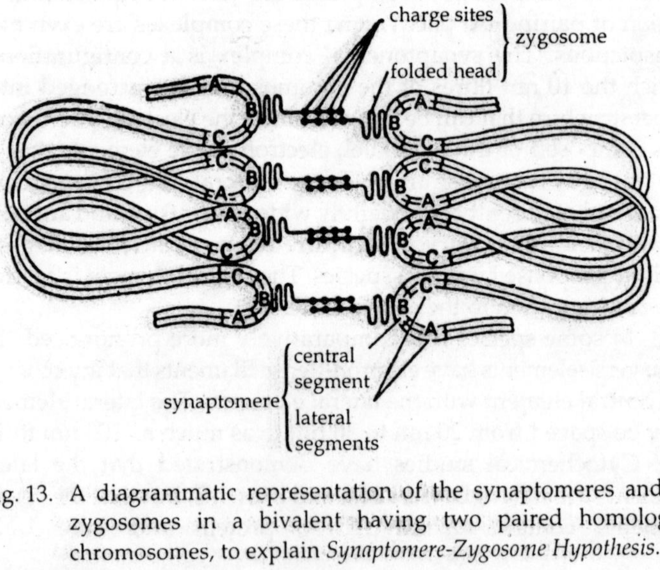

Fig. 13. A diagrammatic representation of the synaptomeres and the zygosomes in a bivalent having two paired homologous chromosomes, to explain *Synaptomere-Zygosome Hypothesis.*

The tail ends of the zygosomes contain charge sites that are represented by four dots. These charges allow the zygosomes to bind laterally with adjacent zygosomes in a lader-like fashion. The exact nature of events that lead to synapsis is still a subject of debate. A strong group of investigators believe that homologous chromosomes are prepared for synaptic pairing by the attachment of their telomeres to so-called 'attachment sites' on the nuclear envelope. This is believed to be followed by attachment of zygosomes to central segments of synaptomeres and to adjacent zygosomes. When each synaptomere has a zygosome, pegs extend from chromosome folds and pairing extends in a zipper like action.

Function of the Synaptonemal Complex

The appearance and disappearance of the synaptonemal complex coincide with the stages of meiosis in which pairing and recombination occur. This has led to the interpretation that they are functionally related. In leptonema, before pairing, single elements of synaptonemal complex are observed. Complete synaptonemal complexes are seen at zygonema in the region of pairing. Several pieces of evidence indicate that the synaptonemal complex is more directly related to the process of recombination. Some evidence, for example, is provided by the action of inhibitors of DNA synthesis.

At meiotic prophase, there is a small amount of DNA synthesis which, if inhibited, can arrest the function of the synaptonemal complexes. The synaptonemal complex has been interpreted as a protein framework that permits the proper alignment of the homologous chromosomes. However, since recombination by crossing over occurs at molecular level, it is necessary to assume that DNA fibres of the paired chromatids should reach the central component of SC within a distance of at least 1.0 μm for the recombination to take place. At diplonema the synaptonemal complex is shed from the bivalents with the exceptions of the regions in which the repelling homologues are held together each by a chiasma. Thus, a chiasma contains a piece of synaptonemal complex that will ultimately disappear and will be replaced by a chromatin bridge.

A Comparison of Mitosis and Meiosis

Now that we have considered both mitosis and meiosis, let us briefly compare the two processes.

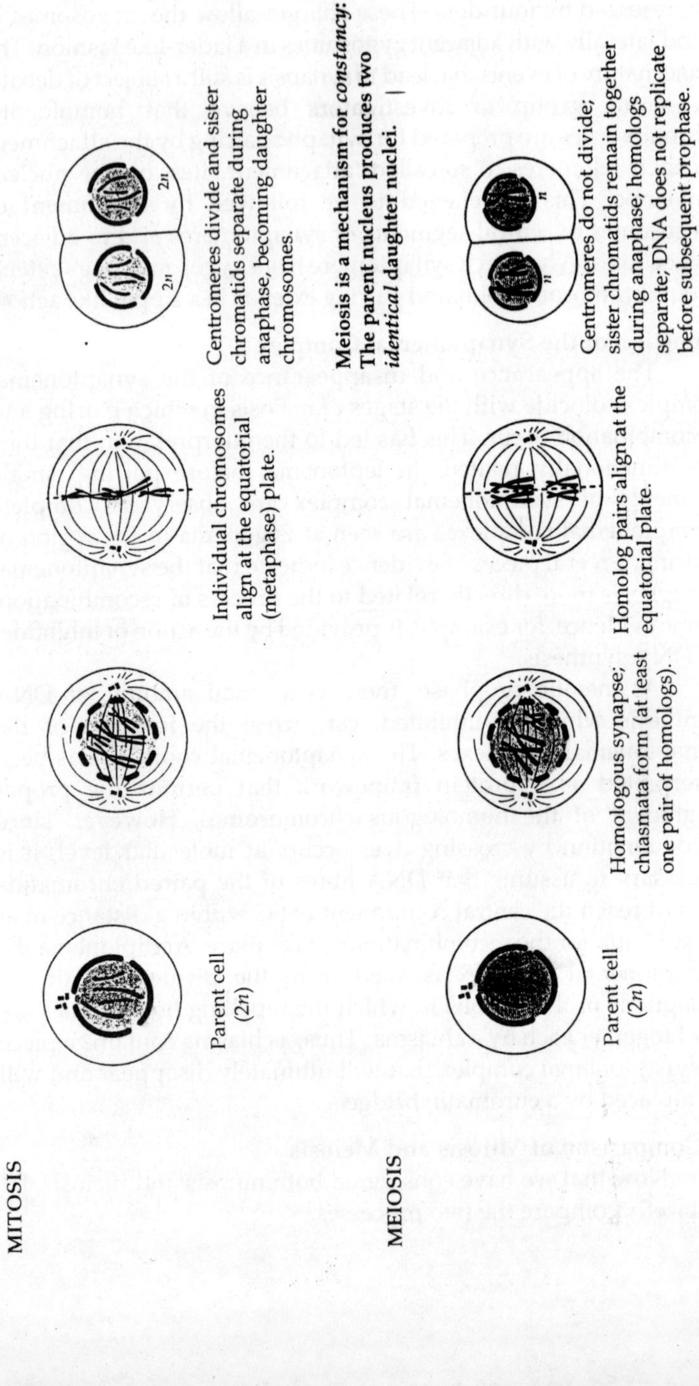

Meiosis is a mechanism for *constancy*: The parent nucleus produces two *identical* daughter nuclei.

Fig. 14.

Mitosis and meiosis are similar in that they are both forms of cell division, incorporating complex mechanisms that ensure that the genetic information is reliably passed to each new cell. But these two methods of division play different roles in the life cycles of organisms.

The outcome of mitosis is simply the production of new cells from previously existing cells. This is the method by which multicellular organisms grow and also by which wornout cells are replaced. In contrast, meiosis is necessary for the reduction of chromosome numbers in the life cycle of organisms that reproduce sexually.

Mitosis may occur in haploid cells, diploid cells, or polyploid cells, and it does not result in a change in chromosome number. On the other hand, meiosis always produces cells with half as many chromosomes as the original cell.

The process of mitosis takes place in one division with four stages. Meiosis requires two divisions: one to separate the homologous chromosomes and one to separate the replicated chromatids.

Finally, the products of the two processes differ. Mitosis produces two new cells, each of which has exactly the same genetic information as the parental cell. In contrast, meiosis produces four cells, each of which has half as many chromosomes as the original parental cell.

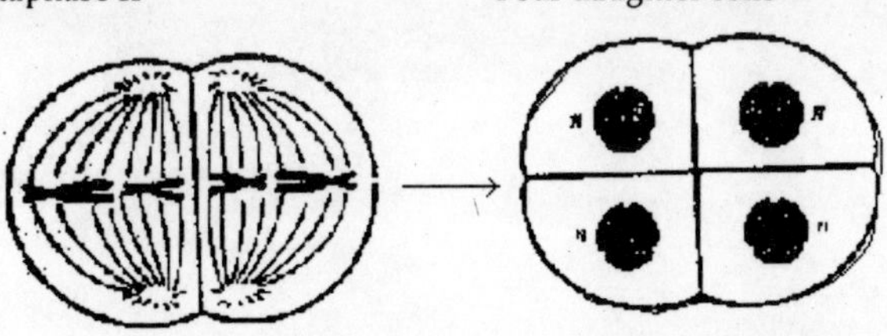

Fig. 15. Mitosis and Meiosis compared.

Meiosis differs from mitosis by synapis and by the failure of the centromeres to separate at the end of metaphase I. Meiosis is a mechanism for diversity: The parent nucleus produces four daughter nuclei, each different from the parent and from its sisters.

3

Interaction of Genes

Genes are the hereditary units and found in the chromosomes. Genes found in the same chromosome are the *non-allelic genes* and those found in the homologoss chromosomes are called the *allelic genes*.

Ordinarily in most populations one allele is found at high frequency at each gene locus. Most of the chromosomes in the members of that population carry that allele, and the phenotype it determines will be the norm for the populations of that species. Such alleles are called *wild-type alleles*. For example, *Drosophila melanogaster* has a wild-type allele determining light grey-brown body colouration with some black and yellow-white areas.

Mutations at a sex-linked locus designated as yellow give rise to alleles that are recessive in expression and give a yellow body and may modify the colour of the bristles. The wild-type allele is designated as y^-. A large number of other alleles are known at this locus. Some of them are listed in Table 1 alongwith the phenotypes produced when they are homozygous in the females and hemizygous in males. All of these alleles are recessive to the y^+ allele. Taken together they form a *multiple allelic series*. Many loci in *Drosophila* and other organisms have multiple alleles that are distinguishable from one another phenotypically because they make a related but slightly different morphological change in the shape of parts or colour of the body. Many alleles, on the other hand, must be distinguished by their physiological or biochemical effects.

A mutant gene in *Neurospora* can cause a nutritional requirement for a specific amino acid or vitamin not required by the wild type. What has happened is that wild-type allele has mutated to a mutant allele that produces a protein without enzymatic activity. As we will see, many different kinds of mutations of a single gene can occur to produce a spectrum of alleles. Thus, the *trp*-3 locus of *Neurospora,* which controls the *synthetase,* has scores of mutant alleles. They cannot all be distinguished by growth tests, because strains carrying these alleles all have a requirement for tryptophan and are not satisfied by any of the tryptophan precursors. Thus they all have approximately the same phenotype. However, it can be shown that they are different at the molecular level and produce proteins that differ slightly from one another and from that produced by the wild type.

Interaction of Allelic Genes

In most allelics series one of the alleles is dominant to all the others. Generally this is the wild-type allele but there are also numerous cases of mutant alleles being dominant. Indeed, many of the known inherited human conditions are dominant to the normal.

Table 1

Some Alleles of the Yellow Locus in *D. Melanogaster*

Allele	*Phenotype*
y^+	Wild type
y	Body yellow, bristles brown with yellow tips
y^2	Body yellow, bristles black
y^{2s}	Darker than y^2
y^3	Tannish body, bristles dark brown
y^{3d}	Yellow body and bristles
y^5	Lethal in male
y^{35a}	Similar to y

Several examples of these kinds of interactions are also found in plants. In the snapdragon, *Antirrhinum majus,* the genes *N* and *P* interact in the formation of the pigments that give colour to flowers. They each have recessive alleles, *n* and *p*. The homozygous *NN PP* plant has the same colour flowers as the heterozygote *Nn Pp,* which in our example here is yellow-orange. The results of a dihybrid cross indicate the genotypes, and the phenotypes expected. The phenotypic ratio is 9 yellow-orange: 4 pure white: 3 ivory. This is because when *n* is homozygous no pigment whatever is formed, and when *p* is homozygous with *N* present, a different pigment colour, ivory, is produced. If it is assumed that these two genes directly control different reactions in the synthesis of the flower pigments, and that their recessive alleles are inactive, then the results can be easily explained.

	NP	*Np*	*nP*	*np*
NP	*NNPP*	*NNPp*	*NnPP*	*NnPp*
Np	*NNPp*	*NNpp*	*NnPp*	*Nnpp*
nP	*NnPP*	*NnPp*	*nnPP*	*nnPp*
np	*NnPp*	*Nnpp*	*nnPp*	*nnpp*

Phenotypic ratio

N—P–Yellow–orange $\frac{9}{16}$

N–pp–Ivory $\frac{3}{16}$

nnP / *nnpp* } White $\frac{4}{16}$

Fig. 1. A dihybrid cross in the snapdragon.

As shown in the figure, *N* controls the formation of a precursor to all the pigments present in the yellow-orange plants. When its recessive allele is homozygous, however, precursor 1 is not formed and hence the flowers have no pigment and are white. When both *N* and *P* are present, all the pigments are present, and this mixture gives the yellow-organge colour. *P* controls the conversion of precursor 5 to the pigments kaempferol and pelargonidin, and when its allele, *p*, is homozygous, these are not formed. Only apeginin and aureusidin are formed, and the flowers are ivory coloured. There are other genes involved, of course, but we have considered only two here so as not to make matters too complicated. *N* and *P* interact in the sense that both must be present to form all these pigments. But if the plant is *nn*, it makes no difference what the allele at the *P* locus is.

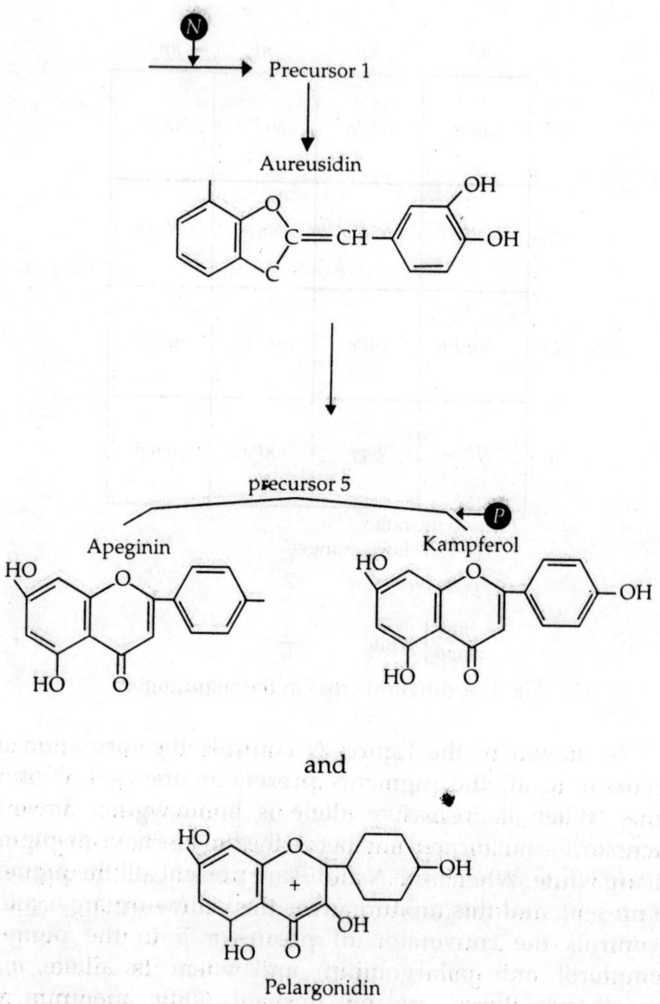

Fig. 2. The possible action of genes *N* and *P* in the snapdragon.

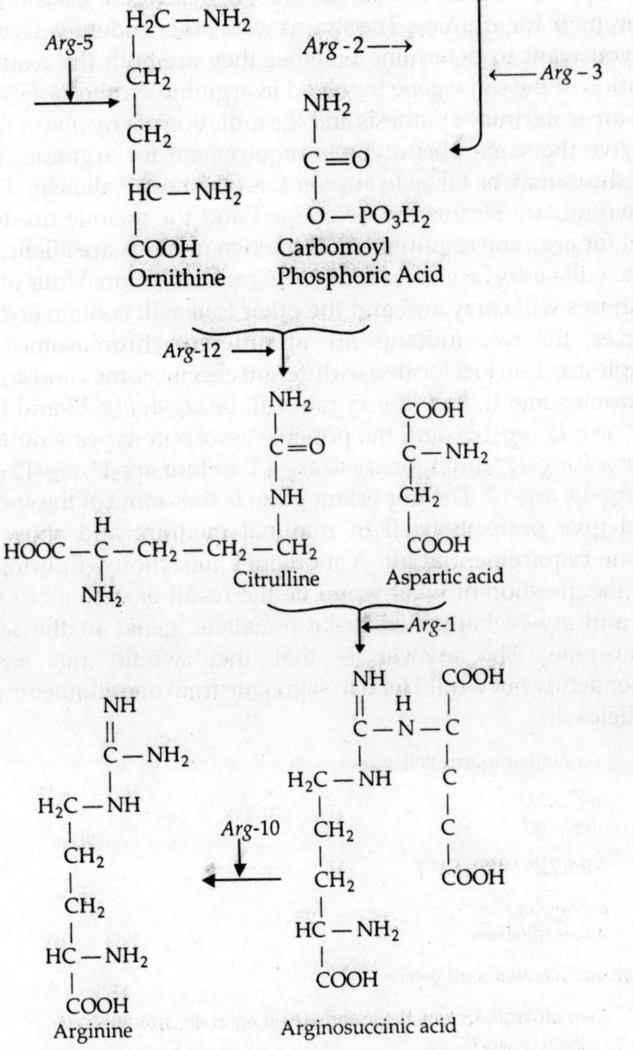

Fig. 3. The biosynthesis of arginine and the genes involved.

Complementary Genes in Haploid Organisms

Suppose you have two cultures of *Neurospora* each with a requirement for arginine. The strains were independently derived, and you want to determine whether they are both the result of mutation of the same gene involved in arginine synthesis. Several steps are in agrinine synthesis and the mutation of any one of them will give the same phenotype, a requirement for arginine. Two procedures may be taken to answer this question of allelism. First, the two mutant strains can be crossed and the meiotic products tested for arginine requirement. If the two mutants are allelic, the zygote will be $arg^x\ arg^y$, and thus (with rare exceptions) four of the ascosposes will carry arg^x and the other four will contain arg^y. If, however, the two mutants are at different chromosomes for example arg-1 on loci located in different chromosome I and *arg*-12 on chromosome II, then the zygote will be *arg-1; arg-12* and four $arg\text{-}1^+$; *arg*-12 $arg\text{-}12^+$, and the possible ascospore types would be four *arg-1*; $arg\text{-}12^+$ and four *arg*-1; $arg\text{-}12^+$ or four $arg\text{-}1^+$; *arg*-12 and four $arg\text{-}1^+$; arg-12. The important point is that some of the spores would give perfectly well in minimal medium and show no arginine requirement at all. A moment's reflection will bring to mind the question of what woud be the result of such a cross in *arg*-1 and *arg*-12 happened to be nonallelic genes in the same chromosome. The answer is that they would not assort independently but would tend to segregate from one another much like alleles do.

(a) Two mutants are allelic

$arg^X \times arg^Y$

arg^X/arg^Y

MEIOSIS WILL GIVE

$arg^X arg^X arg^Y arg^Y$

all arginineless

the action and interaction of genes

(b) Two mutants are nonallelic and are on separate chromosomes

$arg\text{-}1 \times arg\text{-}12$

$$\frac{arg-1 \quad arg-12}{arg-1^+ \quad arg-12^+}$$

MEIOSIS WILL GIVE

arg requiring $\begin{cases} arg\text{-}1\ arg\text{-}12 \\ arg\text{-}1\ arg\text{-}12 \end{cases}$

$$\text{wild types} \quad \begin{cases} arg\text{-}1^{-} \; arg\text{-}12^{+} \\ arg\text{-}1^{-} \; arg\text{-}12^{+} \end{cases}$$

$$\text{or } arg \text{ requiring} \quad \begin{cases} arg\text{-}1 \; arg\text{-}12^{+} \\ arg\text{-}1 \; arg\text{-}12^{+} \\ arg\text{-}1^{-} \; arg\text{-}12 \\ arg\text{-}1^{-} \; arg\text{-}12 \end{cases}$$

Fig. 4. Consequences of allelism and nonallelism.

Nonallelic genes on the same chromosome do recombine and their nonallelism may be established by making a cross and examining the offspring for wild-type recombinants. It is also possible to have rare recombinations between alleles. A second approach to establish whether two mutants are allelic in *Neurospora* is to make heterokaryons. If the two arginine mutants are of the same mating type they will form a heterokaryon. If two different loci are involved, the heterokaryon will grow on minimal medium as the figure shows. This is an example of *nonallelic* or *intergenic complementation*. The test is subject to the same limitations as we have already described for the yellow locus alleles in *Drosophila*. Sometimes the heterokaryon will grow even if both mutations are in the same locus because of intragenic complementation as explained for the yellow locus heterozygote y^2/y^{35a} in *Drosophila*. In this event, a cross between the two mutant strains can settle the question.

If the cross between the two mutants produces only mutant progeny the two mutants are allelic, but if wild types as well as mutant types are produced then the genes are non-allelic. Heterokaryons and heterozygotes can also be made between strains with quite different nutritional requirements. Consider an example from yeast. The life cycle of yeast includes both haploid and diploid vegetative phases. If two haploid strains of yeast with different genetic blocks are of different mating types, they may be crossed. The haploid cells will fuse, and the transient heterokaryon will grow on minimal medium. Soon the nuclei will fuse, and the diploid heterozygote will grow just as well as a *Neurospora* heterokaryon.

Table 2

Some Different Kinds of Dihybrid Mendelian Ratios. Various Types of Interactions may be Deducted by the Phenotypic Ratios Obtained. Study These Ratios and Phenotypic Results Carefully and Make Your Own Deductions as to why the Interactions are Named as they are. For example in case of the gene is obviously acting as an Inhibitor of *B*, which is Required for Colour

<table>
<tr><th rowspan="3">Organism</th><th rowspan="3">Character</th><th rowspan="3">Ratio</th><th colspan="9">Genotype and Number</th><th rowspan="3">Type of Interaction</th></tr>
<tr><th>AABB</th><th>AABb</th><th>AaBB</th><th>AaBb</th><th>AAbb</th><th>Aabb</th><th>aaBB</th><th>aaBb</th><th>aabb</th></tr>
<tr><th>1</th><th>2</th><th>2</th><th>4</th><th>1</th><th>2</th><th>1</th><th>2</th><th>1</th></tr>
<tr><td>1. Peas</td><td>Seed shape and colour</td><td>9:3:3:1</td><td colspan="4">Gray-brown, round</td><td colspan="2">Gray-brown angular</td><td colspan="2">White, round</td><td>White, angular</td><td>None: regular F_2 ratio</td></tr>
<tr><td>2. Chickens</td><td>Comb shape</td><td>9:3:3:1</td><td colspan="4">Walnut</td><td colspan="2">Rose</td><td colspan="2">Pea</td><td>Single</td><td>Regular with interaction</td></tr>
<tr><td>3. Mice</td><td>Colour</td><td>9:3:4</td><td colspan="4">Agouti[a]</td><td colspan="2">Black</td><td colspan="2">White</td><td>Albino</td><td>Complementary (for agouti)</td></tr>
<tr><td>4. Squash</td><td>Colour</td><td>12:3:1</td><td colspan="6">White</td><td colspan="2">Yellow</td><td>Green</td><td>Inhibitor (A) of colour</td></tr>
<tr><td>5. Swine</td><td>Colour</td><td>9:6:1</td><td colspan="4">Red</td><td colspan="4">Sandy</td><td>White</td><td>Complementary (for red)</td></tr>
<tr><td>6. Sweet peas</td><td>Colour</td><td>9:7</td><td colspan="4">Purple</td><td colspan="5">White</td><td>Complementary (for purple)</td></tr>
</table>

Organism	Character	Ratio	*Genotype and Number*									Type of Interaction
			AABB 1	*AABb* 2	*AaBB* 2	*AaBb* 4	*AAbb* 1	*Aabb* 2	*aaBB* 1	*aaBb* 2	*aabb* 1	
7. Chickens	Leg feathers	15:1	Feathered								Featherless	Duplicate genes
8. Chickens	Colour	13:3	White						Coloured		White	Inhibitor of Colour
9. Mice	Colour	10:3:3	White spotted				White		Coloured		White spotted	Complex Interaction
10. Wheat	Kernel colour	1:4:6:4:1	Dark red	Medium dark red		Medium red		Light red	Medium red	Light red	White	None multiple genes

[a]Black hair with a yellow-raddish band, giving gray appearance.

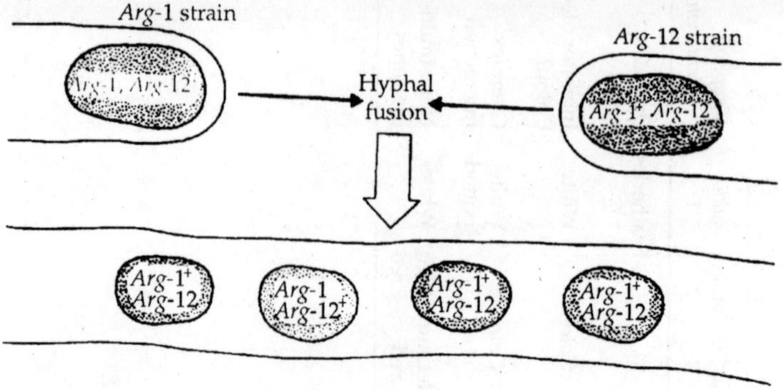

Fig. 5. Heterokaryon with two mutant genes both associated with genetic blocks in arginine synthesis.

Interaction of Non-allelic Genes

It is not necessary that a single character is controlled by one gene only, instead it may be controlled by more than one gene. In such a situation two or more than two genes may interact to give rise to a particular phenotype. For instance, if gene 'A' is responsible for phenotype 'A' and gene 'B' is responsible for phenotype B, both 'A' and 'B' when present together may give rise to a phenotype C. Nevertheless, these genes still obey the principles of segregation and independent assortment. If "A" and "B" are two genes, each dominant over its respective recessive allele "a" and "b", then the interaction will depend upon: (i) the presence of both dominant alleles A and B, (ii) the absence of A, (iii) the absence of B or (iv) the absence of both A and B. As indicated above, the interaction is possible between more than two genes also, but due to the complexity involved in those cases, interaction involving only two genes will be considered in this chapter.

Abbreviated Ratio

On the basis of the four alternatives outlined above, we can give an abbreviated genotypic ratio expected in F_2 i.e., 9 AB: 3Ab: 3aB:, 1ab. In this ratio, AB means those individuals which have both A and B either in homozygous condition or in heterozygous condition (1AaBB: AABB: 2AABb: 2AaBB: 4AaBb). Similarly, Ab means absence of B, but presence of A in homozygous or heterozygous condition (1AAbb: 2Aabb). Likewise, aB includes two genotypes (1aaBB: 2aaBb) and ab will include only one genotype (1aabb).

The abbreviated genotypic ratio is very useful in predicting phenotypic ratio for a particular kind of interaction. If the abbreviated genotype is understood and its use practised, no checkerboard or forked line method will be needed for working out phenotypic ratios. In the following text some cases of interaction of genes will be discussed using checkerboard as well as the abbreviated genotypic ratios. Checkerboards are being used for the convenience of those who find difficulty in using abbreviated genotypic ratios. Otherwise, the checkerboard can be easily dispensed with.

Table 3

Genotypes and Abbreviated genotypic ratio obtained in F_2 from F_1 (AbBb).

Genotypes	*Ratio*	*Abbreviated*	*Ratio*
AABB	1		
AaBB	2	*AB*	9
AABb	2		
AaBb	4		
AAbb	1	*Ab*	3
Aabb	2		
aaBB	1	aB	3
aaBb	2		
aabb	1	ab	1

Lethal Genes

Let us consider first a classical example of a gene the phenotypic manifestation of which is not interfered with by environmental variations but which nevertheless alters the usual Mendelian ratios. The classical Mendelian segregation of ¾ dominants: ¼ recessives occurs when the gametes that carry different alleles of the same gene unite at random and when different gametes and zygotes are equally viable and vigorous. These conditions are not always realised. Not long after the rediscovery of Mendel's work, Cuénot noticed that matings between two mice with yellowish fur produced progenies with yellow and non-yellow individuals in a ratio of 2 : 1.

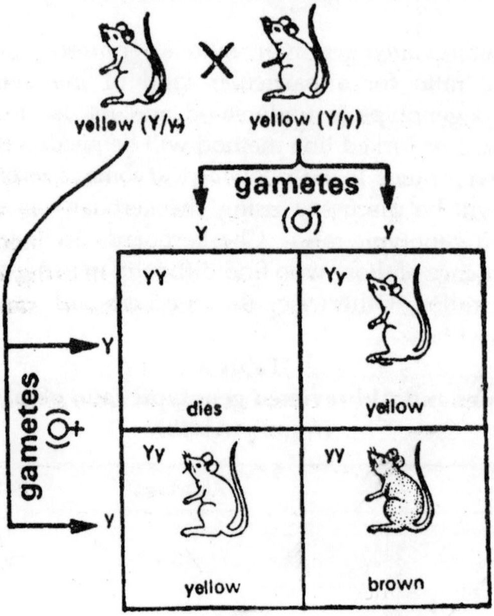

Fig. 6. Segregation for yellow body colour when yellow mice are intercrossed.

It was found that the litters born from matings between yellows are smaller by about one-fourth than litters from yellow × non-yellow. Both these facts are consistent with the hypothesis that the zygotes homozygous for yellow are inviable and die before birth. The hypothesis has been varified by several investigators who found that some embryos in the uterus of a yellow female mated to a yellow male die in an early embryonic stage.

Table 4
Inheritance of the Difference between Yellow and Non-yellow in Mice

Parents	*Offspring*	
	Yellow	*Non-yellow*
Yellow × yellow	2,396	1,235
Yellow × non-yellow	2,378	2,398

The inheritance of yellow may be represented as follows:

A^Y = a dominant gene for yellow fur

a = its recessive allele for non-yellow

Parents:

Yellow A^Ya × Yellow A^Ya

Progeny:

¼A^YA^Y	:	$\frac{2}{4}A^Ya$	:	¼aa
die		yellow		non-yellow

Thus, a mouse homozygous for A^Y dies before birth. The allele A^Y is *lethal*, i.e., produces death, in homozygous condition. When heterozygous, it merely modifies the colour of the fur, which assumes a yellowish tinge. The effects of the allele A^Y on fur colour are dominant over those of a, and its lethal effects are recessive; mice heterozygous for A^Y do not seem to be particularly handicapped by having fur of a different colour. This is an example of some of the effects of one gene's being dominant while others are recessive to those of its allele. Such situations are not rare among lethal as well as non-lethal genes. Dominant and recessive alleles are not sharply distinct caregories.

Lethality in Man

Lethal genes are known in many species of experimental animals and plants. In Drosophila many hundreds of them have been detected, and in man they are also numerous. The gene causing the disease called *juvenile amaurotic idiocy* may serve as an example. Infants homozygous for this gene appear to be normal at birth. However, their eyesight begins to fail between the ages of 4 and 7, resulting gradually in blindness. Deterioration of mental and physical powers leads inexorably to death before or during adolescence. Another recessive lethal gene produces, when homozygous, the so-called *Tay-Sachs disease* or *infantile amaurotic idiocy*, which results in early death of the homozygotes.

The parents of amaurotic idiots are normal people, usually quite unaware of being heterozygous carriers of the recessive genes capable of producing a dreadful disease when homozygous. Denoting the recessive allele responsible for the disease by a and its normal allele by A, the parents of the afflicted children are heterozygotes, Aa.

According to the first law of Mendel, one quarter of their children will be homozygous aa and will die of the disease. Among the surviving children, two-thirds will be the carriers *Aa*, and one-third will be the normal homozygotes, *AA*. It stands to reason that marriages in which only one of the parties is a heterozygote run no risk of producing afflicted children.

Sickle-cell anemia is caused by a gene with a lethal effect when homozygous and with only a slight but discernible effect when heterozygous. The name of the disease refers to a property of the red blood cells of the carriers, which assume abnormal shapes when placed a media deficient in oxygen—under a sealed cover slip of a microscopic preparation, for example, the homozygotes usually die of a fatal anemia before the advent of sexual maturity (although they may occasionally survive). The heterozygotes are healthy people, occasionally showing signs of a mild anemia. Since the heterozygotes are identifiable by means of a microscopic examination of their blood, the birth of children with the fatal anemia could be prevented by avoidance of marriages among the carriers. There is another hereditary disease called *thalassemia* or *microcytemia* which is a somewhat analogous condition, revealed by abnormalities of the red blood cells in the heterozygotes. The gene in homozygous condition causes a fatal anemia (Cooley'anemia). It was thought to occur chiefly among peoples native to countries near the Mediterranean Sea (Italians, Greeks, and Syrians), but is found also in Persia, India and Thailand.

Phenylketonuria

Many forms of idiocy or imbecility are known in man and in some cases the causative gene has been identified. One of these, *phenylketonuric imbecility,* was identified by the discovery of a peculiar reaction of the urine of some inmates of homes for mentally defective children. The urine of certain patients developed a temporary deep bluish-green colour upon addition of a few drops of a 5 per cent solution of ferric chloride. This reaction revealed the presence of phenylpyruvic acid, which is excreted by these patients but metabolized in normal persons.

All persons who continuously excrete this substance have some degree of mental impairment, their muscular reflex reactions are accentuated, and they have a characteristic posture and a dilution of skin and hair pigments. They have a reduced life expectancy and seldom have children. All these characteristics are associated with a defect in metabolism (inability to oxidize phenylalanine to tyrosine) caused by a gene that was thought until recently to have no effect in the heterozygote's parents, which appear perfectly normal. This gene could therefore be classed as a recessive, lethal when homozygous.

Recently, it has been found that parents of phenylketonuric idiots have a slightly different tolerance level for the amino acid phenylalanine when they are fed large quantities of it. They cannot metabolize it as rapidly and efficiently as persons who do not carry a gene for phenylketonuria. Thus a means may be provided for diagnosing such heterozygotes, just as persons carrying the sickling and thalassemia genes are identified by blood tests. This wold be the first step in reducing the frequency of such deleterious genes in human populations. It is probable that many prenatal deaths (stillbirths, abortions) in many are due to lethal genes, but genes of this sort have received much less attention and study than those that cause illness and death after birth. Genes that produce lethal changes in embroys have been more extensively studied in other animals—mice, rats, cats, swine, horses and others. Many genes with lethal or deleterious effects have been identified in the vinegar fly, Drosophila.

Semilethal and Subvital Genes in Man

With the possible exception of sickle-cell anemia, the genes that we have described as lethal to mice and men kill all or nearly all homozygous individuals before they reach sexual maturity. These genes are completely lethals. However, genes are also known which kill some of the afflicted persons but permit others to survive long enough to have families. Such genes are called *semilethal* or *subvital.* The distinction between these categories is not sharp, especially since genes that are lethal under some conditions may permit survival under others, e.g., with special medical treatment.

Genes that kill fewer than 100 per cent but more than 50 per cent of their victims are called semilethals, whereas those that cause less than 50 per cent mortality are known as subvital genes. An example of a dominant semilethal in man is a rare disease called *epiloia.* Its victims are heterozygous carriers of a gene that causes a combination (syndrome) of morbid traits, including severe mental deficiency, abnormal growths on the skin, and tumors in internal organs. Most persons with epiloia die during childhood, but some of them manage to survive, marry, and beget children. Half of these children are expected to inherit the dominant semilethal and to develop epiloia. But most of the cases of epiloia are born to parents both of whom are free of the disease; their appearance is due evidently to mutation of the normal gene to a dominant allele responsible for the disease. The frequency of such mutation is estimated at about ten per million gametes. Surviving carriers of epiloia are so rare in most populations that unions of two carriers have not been observed; it is not known what effects the semilethal would have in homozygous condition.

Retinoblastoma is due to another dominant gene; it causes malignant tumors in the eyes, and almost always leads to death in early childhood. This gene acts as an almost complete lethal. According to Neel and Falls, about forty-nine babies born in the Michigan between 1936 and 1945 developed retinoblastoma, most of them being children of normal parents. These "new" cases were assumed to be due to mutation; if the gene were a complete lethal, all cases of retinoblastoma would be due to recent mutations. Occasionally, however, an infant with retinoblastoma survives the critical phase of the disease and grows to maturity. Consequently, the gene for it, like those for sickle-cell anemia and thalassemia must be classed as a semilethal rather than a complete lethal.

Subvital Effects in Mutants of Drosophila

Semilethal or a subvital condition. Of course, semilethal and subvital genes occur not only in man but also in other animal and plant species, wild and domesticated. Many of the classical mutants of *Drosophila* produce subvital alongwith their more conspicuous effects on the external appearance of the flies. Thus vestigial-winged flies are less sturdy than the normal-winged ones. This can be demonstrated by making counts of the vestigial and normal segregants in the F_2-generation flies from the cross vestigial × normal. When the experiment is carried out under favourable culture conditions (ample food, optimal temperature, etc.), the deviations from the segregation ratio of 3 normal : 1 vestigial are no greater than those expected from sampling errors. But if the cultures are handled roughly, and especially if the flies develop in crowded cultures in which the amount of food is small for the number of larvae that grow on it, the proportion of the vestigial flies may fall appreciably below the theoretically expected 25 per cent. The percentages of vestigial-winged flies observed in cultures in which the total numbers of the flies hatched were between the figures indicated are shown in the following tabulation.

Table 6

Total flies per culture	*Per cent vestigial*
Less than 50	24.9
50–100	25.2
100–200	20.8
More than 200	16.0

If crowded cultures, in which only a part of the eggs deposited by the parent flies develop successfully to produce adult progeny, the mortality among the vestigial homozygotes is significantly greater than among the normals. The vestigial class is significantly deficient in such cultures.

Coat Colour in Horses

The entire coat of some horses, including the mane, tail, muzzle, and lower parts of the legs, is chestnut (sorrel) in colour. Mating of chestnut stallions and mares gives only chestnut foals. Chestnut is due to homozygosis for a recessive gene, *b.* The dominant allele, *B,* of the gene turns chestnut into black. Black horses are either homozygous, *BB,* or heterozygous, *Bb,* for this gene. Another dominant gene, *I,* turns black into bay—a brownish coat with black mane, tail, muzzle, and lower parts of legs. A black horse, then, is always homozygous for the recessive allele, *ii,* in addition to having at least one dominant *B.* But neither *I* nor *i* has any visible effects in the absence of *B*; accordingly, chestnut horses may or may not carry *I.* As a result, horses of these colours may have the following genetic constitutions:

Chestnut	*Black*	*Bay*
bb ii	*BB ii*	*BB II*
bb Ii	*Bb ii*	*BbII*
bb II		*BBIi*
		Bb Ii

The matings black × black produce, then, either only black or black and chestnut foals. Black × chestnut may give black, chestnut, and bay. All three colours may also appear in the matings bay × bay, bay × black, and bay × chestnut. Addition of a further dominant, *D*, transforms the coat into dun (dilute yellow), sand-coloured ("buckskin"), or bluish ("mouse") coat, withu black or chestnut mane, tail, muzzle, lower parts legs, and a stripe on the middle of the back. Chestnut, black and bay horses are, then, homozygous, *dd*, for the recessive allele, *d* (= non-dun).

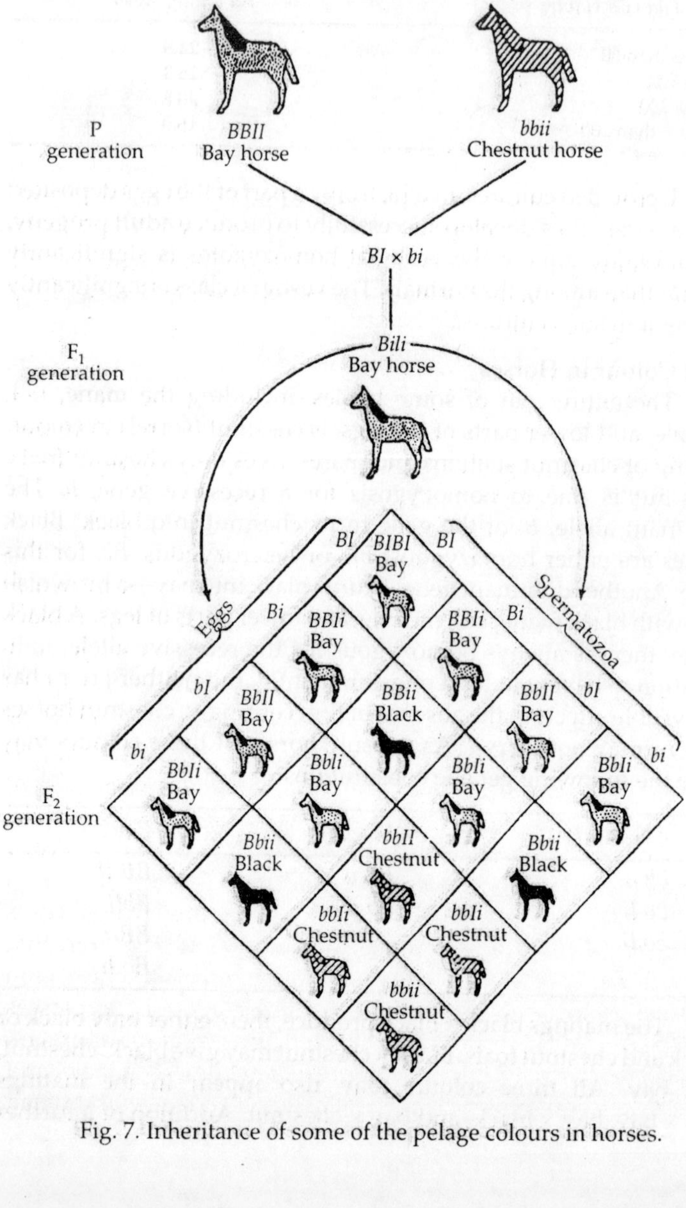

Fig. 7. Inheritance of some of the pelage colours in horses.

The dun colour (*BBIIDD*) is characteristic of the wild horse of Mongolia (*Equus przewalskii*), which is one of the progenitors of the domestic horse. A race of the same species, known as tarpan, lived in the steppes of Eastern Europe until it was finally destroyed a century ago. This race was mouse coloured and probably had the genes *BBiiDD*. Two further genes, *G* and *R*, make the horse respectively gray and roan. A foal homozygous or heterozygous for *G* (*GG* or *Gg*) is born with only a few white hairs scattered in an otherwise chestnut, black, bay or dun coat (depending on the other genes). The proportion of white hairs rapidly increases, however, with age, making the horse first dappled gray and eventually white (with black skin).

The gene for roan (*R*) acts in a different way: it causes the appearance of rather numerous white hairs interspersed with hairs of other colours, the amount of white not greatly increasing with age. There are several further genes the effect of which on the coat colour are not completely understood. A dominant gene cause the horse to be pied with large white areas on the trunk (pied, pieblad, or pinto horses). The presence of white markings on the face and on the lower parts of the legs seems, however, due to a quite different gene, which is recessive to its "normal" allele, causing absence of white markings on these parts. A "white" horse, having a light skin (that is, not an old gray), may be due to extreme development of the "pinto" markings. There is another gene, however, which gives a white coat when homozygous and a characteristic pale ("palomino") coat in heterozygous condition.

Combs in Fowls

A classical case of two genes influencing the same character was discovered about half a century ago by Bateson and Punnett in fowls. Each variety of poultry possesses a characteristic type of comb. The Wyandotte breed, among others, has a comb known as the "rose" comb; Brahmas and some other varieties have a comb known as the "pea" comb; Leghorns and breeds of similar origin have "single" combs. Each of these types can be bred true. Crosses between rose-combed and single-combed varieties showed that rose was dominant over single and that there was a segregation into three-fourths rose and one-fourth single in the F_2. In crosses between pea-combed and single-combed birds, pea was dominant over single, and a 3 : 1 ratio appeared in the F_2. A new and interesting result, however, was obtained when rose was crossed with pea, for the F_1 birds showed a new comb form known as "walnut" previously noted as characteristic of the Malay breed of fowls, a race unrelated to those from which the new walnut comb was obtained.

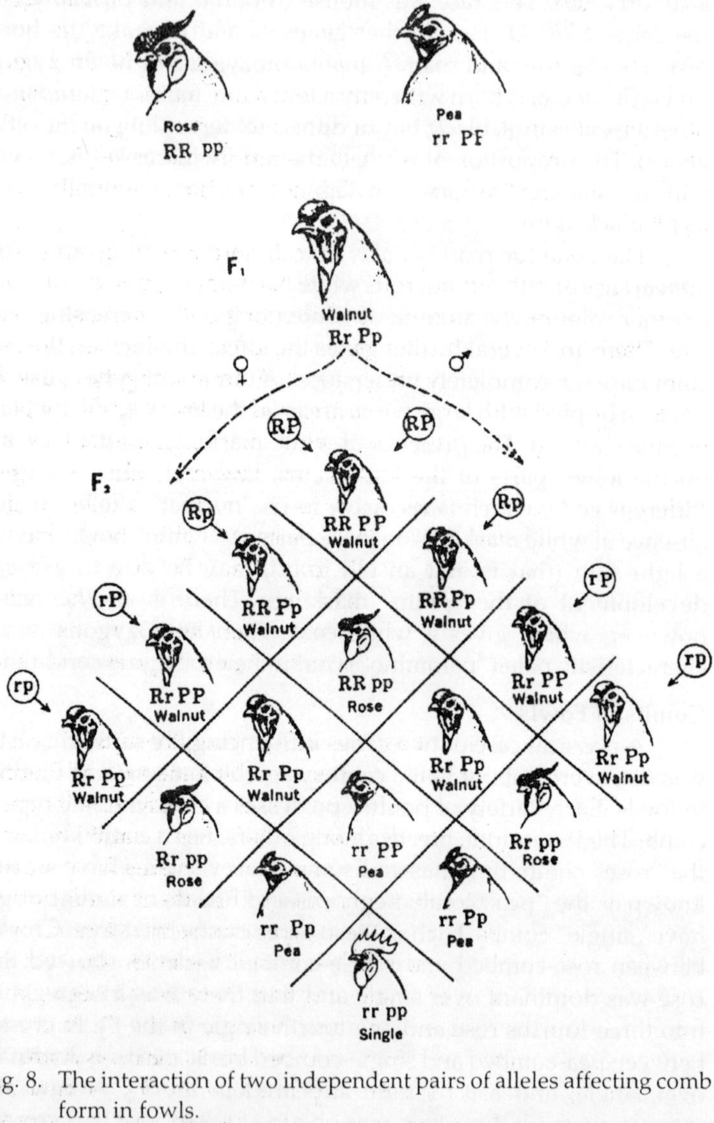

Fig. 8. The interaction of two independent pairs of alleles affecting comb form in fowls.

When the F_1 walnut-combed birds were bred together, a still more remarkable result was manifest, for in the F_2 generation there appeared not only walnut-, rose- and pea-combed fowls but single-combed ones as well. These types occurred in the proportions 9/16 walnut, 3/16 rose, 3/16 pea, 1/16 single. This ratio is expected in F_2 from a cross of parents differing in two genes. The doubly dominant class in F_2 was walnut, whereas he numbers of singles obtained indicated that this type contained both the recessive genes involved, a conclusion supported by the fact that the F_2, singles, when bred together, produced only single-combed progeny in subsequent generations. The walnut comb depends on the presence of two dominant genes, *R* and *P*. One of these genes alone *(R)* produces the rose comb; the other alone (*P*) produces the pea comb. The combination of the recessive alleles of these genes produces the single type of comb, *rp*.

The mode of inheritance of the genes for rose and pea does not differ at all from the usual Mendelian scheme. The differences that distinguish this and similar cases from ordinary dihybrid inheritance are that: (1) the F_1 resembles neither parent, and (2) apparently novel characters appear in F_2. One of these new characters (walnut comb) evidently results from an interaction between two independently inherited dominant genes, and the other (single comb) results from the interaction of their two recessive alleles. These peculiarities are due not to a new method of inheritance but simply to the circumstance that both genes involved happen to express themselves in the same part of the organism, in his case in the comb.

Complementary Genes for Colours in Corn

Indian corn (maize) occurs in a number of varieties, some of which have colourless seeds, others seeds of a purple colour (due to the presence of a purple substance known as *anthocyanin* in the aleurone layer of the seeds), and still others seeds of various other colours. Some varieties have green leaves, stems, and husks, due to the presence of the common plant pigment, chlorophyll. Others have, in addition, a brownish or a purple anthocyanin pigment in these parts. The crux of the whole matter is that in order to develop the purple anthocyanin pigmentation the plant must carry the dominant alleles not of one but of several genes: plants homozygous for the recessive alleles of any one of these genes have no anthocyanin, at least in the seeds and therefore have colourless seeds. To state it differently, the purple anthocyanin colour is elaborated owing to interaction of the dominant alleles of several genes; the action of these dominant alleles is complementary: that is, each allele controls a step essential for the production of the pigment. The absence of any one of them throws the whole anthocyanin-forming machinery in the plant out of order.

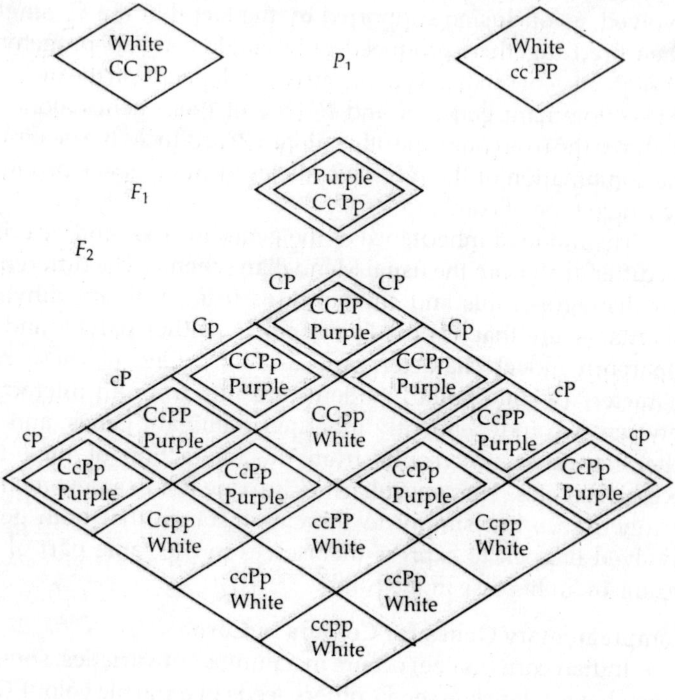

Fig. 9. The 9:7 ratio. Checkerboard showing the expected composition of the F_2 from a cross of two white-flowered sweet peas which produces all purple-flowered palnts in the *F*.

A true-breeding variety of maize with purple seeds and with purple pigment in the body of the plant is homozygous for at least three dominant genes, denoted *A*, *C*, and *R*. The genetic "formula" of such a plant is, accordingly, *AA CC RR*. The recessive allele *a*, in the presence of the dominant *C* and *R*, gives colourless seeds but a plant of a brownish-green colour, *aa CC RR*. The recessive alleles *c* and *r* give colourless seeds and green plants, so that *AA cc RR* and *AA CC rr* genotypes produce plants devoid of purple or brown pigments. The simplest results are obtained when plants with coloured seeds, *AA CC RR*, are crossed to anyone of the three kinds with colourless seeds, *aa CC RR*, *AA cc RR*, or *AA CC rr*. The purple colour behaves as a dominant, and the F_1 seeds are purple. In F_2 the classical segregation in 3 purple: 1 colourless ratios are observed.

Suppose, however, that a colourless-seeded maize plant lacking the gene *A* (i.e., of the composition *aa CC RR*) is crossed with another plant that is colourless but of the composition *AA cc RR*. The f_1 hybrid seeds will have all three colour genes, *Aa Cc RR*, and will be purple in colour. Because the genes *A*, *C* and *R* are complementary, the F_1 hybrids are unlike either parent. The same will happen if a colourless *aa CC RR* is crossed to a colourless *AA CC rr* or a colourless *AA cc RR* is crossed to a colourless *AA CC rr*. The resulting F_1 hybrid seeds will all be purple.

The 9 : 7 Ratio

The f_2 generation from all these crosses will consist of purple and colourless seeds in a ratio of 9 : 7. This ratio illustrates the outcome of the cross that first revealed the complementary interaction of two independent pairs of alleles. Each parent in this cross was from a *different* true-breeding, white variety of sweet pea. The interpretation was that in each white variety, failure to form pigment was due to a different recessive gene and that the presence of both dominant alleles was necessary for the completion of the chain of biochemical processes leading to the formation of a coloured substance. Applying this interpretation to seed colour in corn, we can assume that all seeds that inherit the dominant alleles of the two colour genes, either in homozygous or heterozygous condition, will be purple; the proportion of such seeds will be nine-sixteenths. All seeds that lack the dominant alleles of at least one of the colour genes will be colourless; there will be seven-sixteenths of such seeds.

The validity of this interpretation can easily be tested by a number of experiments of which the outcomes are predictable. For example, we may make the assumption that among the purple seeds there must exist four different genotypes. One-ninth of the purple seeds are homozygous, *AA CC RR*, and the plants grown from them will breed true, i.e., will produce only purple F_2 seeds if selfed or crossed to each other. Four-ninths are double heterozygotes *Aa Cc RR;* they will again give a ratio of 9 purple: 7 colourless seeds if selfed or crossed to each other. Two-ninths are heterozygous for just one colour gene, *Aa CC RR*, and two-ninths heterozygous for another colour gene, *AA Cc RR*. These will give, when selfed or intercrossed, segregation ratios of 3 purple: 1 colourless.

Genetic "Formulas"

In this explanation, only the genes *A-a* and *C-c* need be indicated, and *R* may be omitted, although we know that its presence and the presence of still other genes is necessary if purple pigment is to be formed. Such an omission is customary, for otherwise the genetic "formulas" would be confusingly complex. One includes in a formula the genes that are known to be segregating in a cross and disregards the rest. But there are crosses in which more than two genes must be taken into account. Thus, a colourless-seeded maize of the composition *aa CC RR* can be crossed with a colourless *AA cc rr*, or a colourless *aa cc RR* with a coourless *AA CC rr*, or a purple-seeded *AA CC RR* with a triply recessive colourless *aa cc rr*. In all these crosses, the F_1 hybrids have purple seeds and have identical genotypes, *Aa Cc Rr*. To predict the result obtained in the F_2 generation of these crosses, a checkerboard diagram should be constructed having 64 cells (the trihybrids form, eight kinds of gametes). There will be 27 out of 64 genotypes that will contain all three colour genes, and these will produce purple-coloured seeds. The remainder, 37 out of 64, will lack one, two, or all three dominant alleles of the colour genes and will consequently produce colourless seeds. Such a 37 : 27 ratio has, indeed, been observed in experiments.

Reversion

The inheritance of seed colour in maize and similar situations in other organisms suggest an explanation for the numerous instances among domesticated animals and plants in which crosses between true-breeding varieties produce progeny resembling a remote ancestor more than they do either parent. Plant and animal breeders had noted these peculiar "throwbacks," "atavisms," or "reversions," but in the absence of any satisfactory explanation they regarded reversion as the expression of some mysterious force that caused the retention and subsequent reappearance of a remote ancestral trait.

The reappearance of an old trait is usually due to the reunion of the two or more genes, necessary for its production, which had become separated in the history of the species. Thus, in maize, it is plain from the experiment cited above that purple seed colour depends on at least three genes and that colour less seeds result when any of them is lost. It is easy to imagine that one colourless variety arose when, in the purple type, a mutation occurred from *C* to *c*, and the second arose when *R* changed to *r*. Thus the two elements necessary for purple colour became separated into two different strains. When these strains were crossed, the two genes were reunited, and the primitive, or "reversionary," colour appeared.

Coat Colour in Rodents (The 9 : 3 : 4 Ratio)

A similar but more complex case of gene interaction and reversion has been worked out in breeding experiments with "fancy" varieties of the common house mouse, in which a number of genes have been found to interact in producing what appears to be a simple character. The ancestral coat colour of this species is that of the ordinary wild mouse. Upon close examinaton, this colouration is found to be true to the presence of two pigments in the fur. The individual hairs for the most part black with a narrow yellow band near the tip.

The underside of the animal is usually much lighter, the hairs being cream or yellow, with some black or gray at the base. This inconspicuous and hence protective colouration, which is known as the *agouti* pattern, characterizes many of the wild rodents, such as the Norway rat, the meadow vole, the guinea pig, the gray squirrel, and many others. A number of variations that have taken place in this wild gray, or agouti, coat coloration have been preserved under domestication and have given rise to many colour varieties of mice. The commonest and most familiar variation is the albino, in which the coat is white and the eyes are pink because of the absence of pigment from the iris. Albinos always breed true, and this variation has been found to behave as a simple recessive to any colour.

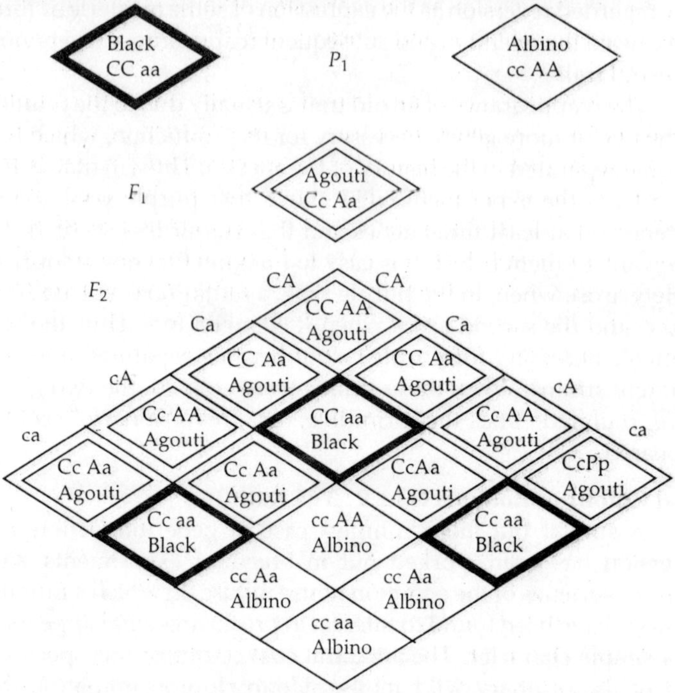

Fig. 10. The 9: 3: 4 Checkerboard showing the expected composition of the F_2 from a cross of black and albino mice which produces all agouti animals (wild-type) in the F_1.

Another variation in coat colour arose through the disappearance of the yellow pigment from the agouti pattern, leaving the fur solid black. Black is recessive to the wild gray type and breeds true. When black mice are crossed with ordinary albinos, the progeny are usually all agouti like the wild type. When these F_1 agoutis are inbred, their progeny consist, on the average, of nine-sixteenths agouti animals, three-sixteenths black and four-sixteenths albino. This, like the 9 : 7 ratio encountered in maize, indicates a difference of two genes. Here, however, the last two terms of the ordinary 9 : 3 : 3 : 1 ratio have been added together, indicating that two of the ordinarily different classes of the dihybrid F_2 zygotes cannot be distinguished. The results are explained on the assumption that the parents differ in: (1) a gene, C, necessary for the development of any colour, which the black mice contain but which is lacking in the albinos: and in (2) a gene for the agouti pattern, *A*, which results in a banding of the black hairs with yellow. Since the black mice cannot contain this gene *A* (if they did, they would appear agouti), it must have come from the albino parent, where, in the absence of the ability to develop any colour at all, it could have no visible expression. The recombination of these two genes, one for colour and the other for the agouti pattern, reconstitutes the genotype of the wild mouse, and a "reversionary" type results. Reversions also occur in other ways, as by return of a mutant gene to its original form (reverse mutation).

Epistasis

Owing to dominance, the presence of the recessive allele in a heterozygote may be obscured or hidden. It sometimes happens that when two different genes, which are not alleles, both affect

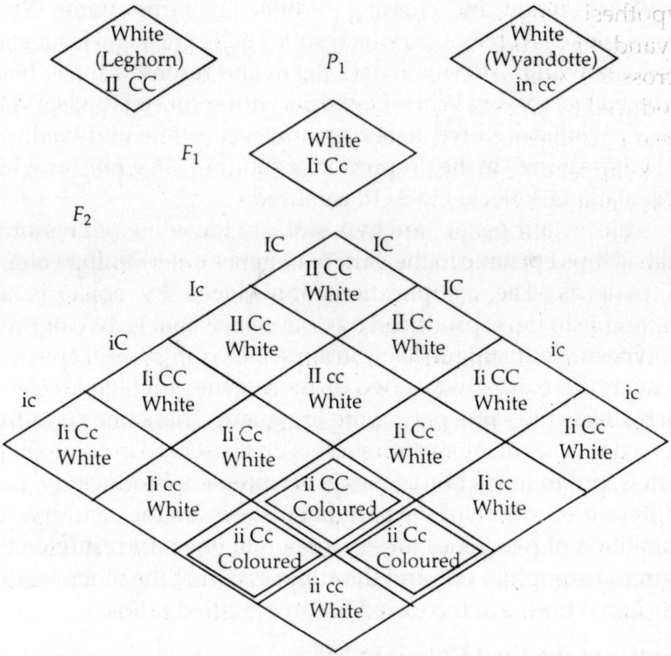

Fig. 11. The 13:3 ratio. Checkerboard showing the expected composition of the F_2 from a cross between two varieties of fowls, one with dominant white plumage and one with recessive white plumage.

the same part or trait of the organism, the expression of one covers up or hides the expression of the other. A gene that masks or prevents the expression of another is said to be epistatic to it, and the gene that is hidden is said to be hypostatic.

The white plumage of White Leghorn fowls is almost completely dominant over the coloured plumage of black, barred, or other coloured varieties. The white plumage of some other white varieties, however, such as White Wyandottes or White Plymouth Rocks, has been found to be recessive to coloured plumage and to be due to a gene distinct from that which produces the white of Leghorns.

Experiments show that White Leghorns contain a colour gene and with it a gene that inhibits its expression. They are genetically coloured birds that are unable to develop their true colour. Denoting such an inhibitor by *I* and the colour gene by *C*, the White Leghorn is *IICC* and the White Wyandotte is *iicc*. A test of this hypothesis made by crossing White Leghorns with White Wyandottes produces a curious result. The F_1 chickens from such a cross are white with small, dark flecks and resemble the F_1 birds produced by crossing White Leghorns with coloured fowls. When these F_1 whites are bred together, however, white and coloured chicks appear in F_2 in the proportion of about 13/16 white (or white with small, dark flecks) to 3/16 coloured.

The "colour genes" are hypostatic to the white of Leghorns, which is thus epistatic to the variety of genes determining colours and patterns. The complications introduced by epistasis are comparable to those produced by dominance; that is, two or more genotypes are indistinguishable in appearance. In cases of epistasis, however, there are always two or more genes involved, each of which affects the same part of the organism. This same condition occurred in cases of interaction such as that observed in comb shape in fowls, but in fowls both genes are expressed, producing a new or different condition of the part. In epistasis, on the contrary, the competition of two genes for expression in one part results in the apparent triumph of one and the suppression of the other, so that the original traits are recovered but in modified ratios.

Analysis of the Coat Colour in Mice

A thorough study of the variations in a group of related characters in any organism will usually reveal an intricate series of interactions between the component genes. As an example we shall choose the house mouse. As we have just seen, *C* is the fundamental colour gene, necessary for the production of any pigment in the coat. Another gene, *A*, determines the development of the agouti pattern. Its recessive allele, *a*, is present in the nonagouti mice, such as blacks or browns. Still another, *B*, governs the develate, *b*.

Many varieties are spotted with white in a blotched, or piebald, pattern, and such mice contain a gene, *s*, which is recessive to self-, or solid, colour, *S*. Another gene, *d*, brings about a clumping of the black and brown pigment granules in the hairs and makes these colours appear faded, or dilutc, as opposed to the normal, fully pigmented form, *D*. Another gene reduces the amount of black and brown pigment in the fur, giving it a pale, washed-out appearance, and also reduces the pigment in the iris, making the eyes appear reddish or pink like the eyes of albinos. This gene, which is called pink eye (*p*) from its most noticeable effect, is recessive to the normal dark-eyed, intense-coloured condition, *P*. These genes all segregate sharply and may occur in any combination. There are also several other genes affecting coat colour which, in order to simplify the problem, will not be discussed here. Some of these combinations result in characters that are distinctive and have been given names of their own. Thus the non-black agoutis are called "cinnamon" or brown agouti: the dilute blacks, "blue"; the dilute browns, "silver fawn"; and so on.

Table 7

Interaction of Genes for Coat Colour in Mice

Genes					*Gametic formula*	*Phenotype*
			P	*S*	*CABDPS*	Wild-type agouti
		D		*s*	*CABDPs*	Spotted agouti
			P	*S*	*CABDpS*	Pink-eyed agouti
	B			*s*	*CABDps*	Pink-eyed, spotted agouti
			P	*S*	*CABdPS*	Dilute agouti
			P	*s*	*CABdPs*	*spotted, dilute agouti*
			P	*S*	*CABdpS*	*Pink-eyed, dilute agouti*
A				s	*CABdps*	Pink-eyed, spotted, dilute agouti
			P	*S*	*CABdpS*	Cinnamon
		D		*s*	*CABDPs*	Spotted cinnamon

			P	*S*	*CAbDps*	Pink-eyed, spotted cinnamon
	b			*s*	*CAbDpS*	Pink-eyed, cinnamon
			P	*S*	*CAbdPS*	Dilute cinnamon
		d		*s*	*CAbdPs*	Spotted, dilute cinnamon
			P	*S*	*CAbdpS*	Pink-eyed, dilute cinnamon
				s	*CAbdps*	Pink-eyed, spotted, dilute cinnamon
C			*P*	*S*	*CaBDPS*	Black
		D		*s*	*CaBDPs*	Spotted black
			P	*S*	*CaBDpS*	Pink-eyed black
		d		*s*	*CaBDps*	Pink-eyed, spotted black
					CaBDPS	Dilute Black
					CaBdPs	Spotted, dilute black
					CaBdpS	Pink-eyed dilute black
					CaBdps	Pink-eyed spotted diluted black
					CaBDPS	Dilute Black
					CaBdpS	Spotted, dilute black
			P	*S*	*CabDPS*	Brown
		D		*s*	*CabDPs*	Spotted brown
			P	*S*	*CabDpS*	Pink-eyed brown
	b			*s*	*CabDps*	Pink-eyed, spotted brown
			P	*S*	*CabDpS*	Dilute brown
		d		*s*	*CabdPs*	Spotted, dilute brown
			P	*S*	*CabdpS*	Pink-eyed, dilute brown
				s	*Cabdps*	Pink-eyed, spotted, dilute brown
c	with any other genes				*c*...........	Albino, may be of any of 32 genotypes above

At any time all the dominant the wild type may be reconstituted by bringing into combination alleles of the genes responsible for these new types. In fact, the wild coat colour itself is found to depend on the presence and interaction of all of the genes named. Thus in order to produce the agouti pattern there must be present the genes for colour (*C*), agouti *(A)*, black (*B*), dark eye (*P*), dense colour (*D*), and solid colour (*S*). With regard only to these genes the genotype of the wild mouse may be written *AA BB CC DD PP SS*. These genes all show essentially complete dominance, so that their heterozygous condition will give the same result as is produced by the homozygous form here given. Thus an animal with the genotype *Aa Bb Cc Dd Pp Ss* would also be agouti in appearance. The genes named do not include all that are known, nor is it believable be that more than a small sample of the genes affecting coat colour in my has been studies. If four knowledge of the subject were complete, it is probable that the list of genes necessary for the production of the agouti pattern would be much longer and that the letters of the alphabet would be exhausted in the attempt to write the genotype of the wild mouse. Here, then, it is clear and convincing example of gene interaction. In order that the apparently simple pattern characteristic of wild house mice may be developed, there must be present at least six genes (probably many more) each of which has a definite effect on coat colour. If any single gene is missing or changed, a coat pattern differing more or less widely from the wild type results.

Gene Interaction in Cats

The work of geneticists has disclosed numerous examples of gene interaction in many species of animals and plants. We shall certainly not attempt to review all of them here, but it may be useful to describe briefly some further examples to show how great is the variety of the situations encountered and at the same time how simply they are explained in terms of the theory of the gene. The validity and importance of a scientific theory are attested by its capacity to provide an intellectually satisfying account of many facts of nature and to help one predict the results of observations and experiments not yet carried out. It has already been pointed out that the agouti colour pattern occurs not only in the house mouse but in other rodents as well.

Other genes having effects similar to or identical with those described for the mouse occur also in some of these other species of rodents. In fact, more or less parallel situations are known in remotely related mammals as well. Thus, in domestic cats there is a gene one allele of which gives, when homozygous, a yellow coat colour, whereas the other allele, when homozygous gives a black colour. But in contrast to mice, this gene is carried in cats in the sex determining X chromosome. In the heterozygous female cats, neither allele is dominant and instead a "tortoise-shell" coat, consisting of patches of black and yellow fur, results. Males cannot normally be heterozygous for genes carried in the X chromosomes, and for this reason male cats are either yellow or black but not "tortoise-shell." Another gene, possibly, parallel to the agouti gene in rodents, gives in cats the "tabby" coat colour, consisting of alternating darker and lighter stripes. This is the colour pattern characteristic of the remote wild ancestors and some living relatives of domestic cats. A recessive gene gives the "Maltese dilution," which transforms black into a smoky gray, tabby into a gray with traces of the tabby pattern, and yellow into a pale sandy coat. The large irregular white spots are due to a dominant gene for spotting.

Albinism and Spotting in Man

In man, there exists a gene that acts like the fundamental colour gene described above for mice. A recessive allele of this gene gives, when homozygous, the albinotic condition described on page 60. The dominant allele of this gene allows pigmentation to develop, and several further genes control its amount, thus causing variations in skin colours ranging from that of the white race to that of the black. There exist also several genes that increase the pigmentation of the hair from blond to black and the eye colour from blue to brown or black.

The interrelations between the genes that increase the pigmentation of the skin, hair, and eyes are not well understood. It is probable that some of these genes affect the pigmentation of all these parts, whereas others act on the hair without changing the eye colour or on the eyes without changing the hair colour. A dominant gene that is rare in human populations causes the skin to be irregularly spotted or pied with darker and lighter areas. The effects of this gene are especially striking in human races that normally have a darky pigmented skin. The analogy between this human gene and the spotting genes in mice, cats, horses, and other mammals is rather striking.

Green Colour in Corn

In corn (maize), many genes affecting the development of the green chlorophyll pigment are known. The normal green colour of maize results from the combined action of the normal alleles of these genes. If one of these genes mutates to a recessive allele, the seedlings are *albinotic* (virtually without chlorophyll); unable to carry on photosynthesis, they soon die. There are at least 15 different genes, susceptible to this type of change, which means that there are at least 15 genetically different types of albinos. In another group of recessive mutants, known as lutescents, only yellow pigment develops; in another group are more than twenty recessive genes, each of which, when homozygous produces virescent seedlings that are albinotic but eventually develop enough chlorophyll to keep them alive.

Other recessive genes are responsible for pale-green colour (10 known), zebra striping (4 known), piebald spotting (4 known), golden colour (4 known), yellow-green colour (3 known), yellow striping (2 known), fine white striping (3 known) and many other modifications of chlorophyll composition and arrangement. It must be true, therefore, that the normal development of chlorophyll in maize depends in interaction among at least 75 different genes. If any one of these changes by mutation of the recessive allele from the normal one, some essential step in the process fails and the chlorophyll is absent or deficient in some way. The results of the analysis of such cases as those described above suggest that many of the characters of organisms are the end products of long chainlike series of related steps, $a \rightarrow b \rightarrow c \rightarrow d \rightarrow n$. Separate genes seem to affect separate steps so that if the gene affecting steps *b* does not perform its task, then *c* and all later steps that depend upon it cannot take place and the character, such as normal chlorophyll, cannot appear. Whatever the means by which the genes interact, it may be accepted as a general rule that the hereditary characters of a plant or animal depend upon the balanced cooperation during development of a large number of genes.

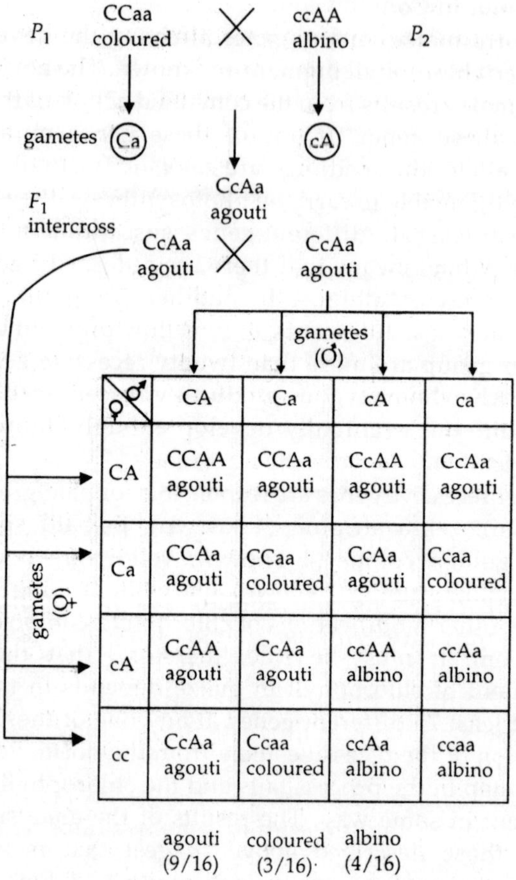

Fig. 12. Checkerboard derived from a cross between coloured and albino mice showing epistasis.

Duplicate Genes

A modified ratio 15 : 1 (derived from standard ratio 9 : 3 : 3 : 1) was observed by G.H. Shull from his studies on the plant known as Shepherd's purse belonging to the genus *Capsella*. Two kinds of phenotypes with respect to fruit shape were known—(i) triangular capsules and (ii) top shaped capsules. When races with these phenotypes were crossed, in F_1 generation only triangular capsules were observed. When such F_1 individuals with triangular capsules were intercrossed among themselves, in F_2 progeny, plants with triangular capsules and top shaped capsules were obtained in 15 : 1 ratio. Obviously the top shaled capsule results from double recessive genotype. If 'A' and 'B' are two genes, top shaped capsules will be obtained on plants with the genotype aabb. Plants with the triangular capsules can be AABB, AAbb, or aaBB and other genotypes with heterozygosity. It shows that even a single dominant gene is enough to give rise to triangular capsules. Presence of both dominants has the same effect.

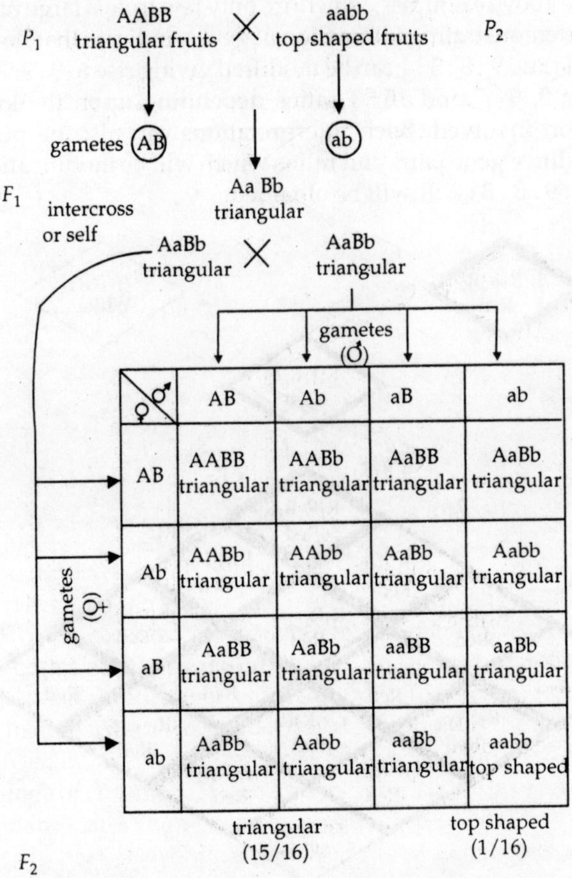

Fig. 13. Checkerboard derived from a cross between two varieties of Shepherd's purse, one having triangular and the other having top shaped capsules (fruits).

Table 8
Abbreviated Genotype Ratio and Phenotypic Ratio Obtained in F_2 in a Cross

	AB	Ab	aB		ab
Abbreviated genotypic ratio	9 :	3 :	3	:	1
Phenotypic ratio	triangular 15			:	top shaped 1

The above examples, which are only few from a large number of them demonstrating interaction of genes, indicate that the basic dihybrid ratio 9 : 3 : 3 : 1 can be modified giving rise to 9 : 3 : 4 : 12 : 3 : 1, 13 : 3, 9 : 7 and 15 : 1 ratios depending upon the kind of interaction involved. Such interpretations can also be possible between three gene pairs and ratios which will be modifications of 27 : 9 : 9 : 9 : 3 : 3 : 3 : 1 will be obtained.

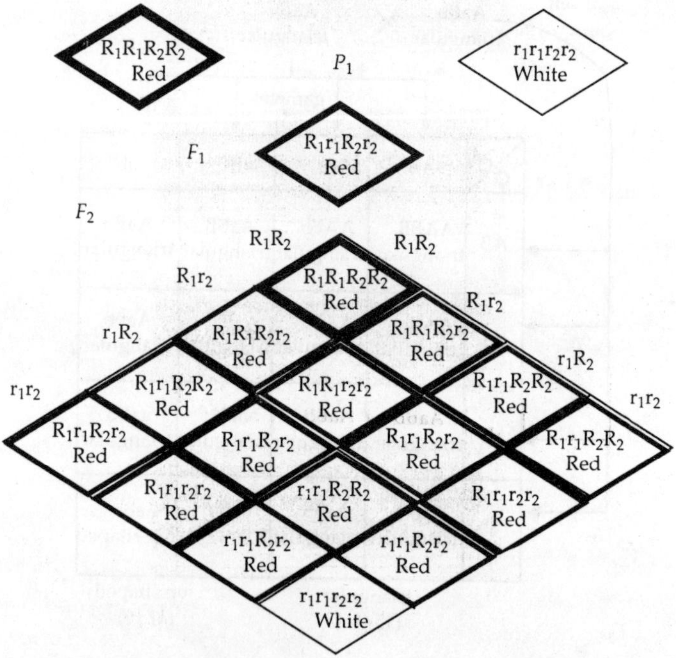

Fig. 14. Diagram showing the result of a cross between a red-kerneled and a white-kerneled wheat, where the red colour is due to the operation of either or both of two genes, R_1 and R_2. The intensity of the red colour is indicated by the density of the crosshatching.

Quantitative and Qualitative Traits

Obviously no sharp dividing line can be drawn between these two kinds of traits. Mendel himself crossed tall and short peas and showed that the difference in size was very simply inherited and caused by just one pair of alleles of a single gene. The tall and the short plants were so distinct that no measurements were needed. It is not precisely the "quantitative" nature of these variations which sets them apart from "qualitative" ones, but the fact that some variations are gradual or continuous so that clear-cut classes can be distinguished only with difficulty or not at all. Thus there occur all kinds of skin colours intermediate between that in nonalbino whites and that in the darkest Negroes. Moreover, quantitative traits are often susceptible to environmental modification. The skin colour of one person can be lighter or darker depending upon the amount of exposure to sunlight.

Animals good and plants grow larger and heavier with abundant food and on good soil.

Polygenes, the genes that bring about heritable variations in quantitative traits, are not a separate category of genes. They are simply genes the different alleles of which produce small phenotypic differences, of about the same order of magnitude as the differences caused by usual environmental fluctuations. Polygenic inheritance is not a separate kind of heredity; it is heredity that must be studied with the aid of statistical techniques rather than by simply counting the numbers of individuals belonging to a few easily distinguishable classes.

Seed Colour in Wheat

Nelsson-Ehle crossed several varieties of wheat, some of which had red and others white kernels. Red is incompletely dominant over white, and the heterozygotes may have somewhat lighter red kernels than the homozygotes. In some crosses of red and white a simple ratio of 3 red : 1 white is found among the F_2 hybrids, indicating a single gene difference. In other crosses a ratio of 15 red : 1 white is found in F_2. The conclusion is obvious (and verified by a study of F_2 generations raised from the red seeds of the F_2) that there are two pairs of genes for red, either of which can produce red kernels. These genes are duplicates of each other.

The checkerboard shows the genotypes of the F_2 plants with the numbers of red genes (capital R letters) possessed by each genotype. Only 1 of the 15 reds has all four genes for red and has dark-red seeds; 4 classes have three red genes; 6 have two; 4 have one red gene and have pale red seeds.

In still other crosses, 63 out of 64 of the F_2 wheat kernels were red and only 1 to 64 was white, a condition which suggested that three genes each capable of producing the red colour were involved. If the red parent is represented by $R_1R_1R_2R_2R_3R_3$ and the white by $r_1r_1r_2r_2r_3r_3$, the F_1 generation has three colour genes, $R_1r_1R_2r_2R_3r_3$. The genotypes that appear in the F_2 generation can easily be predicted by constructing a checkerboard with 64 cells. About 1 in 64 individuals will be like the red grandparent and will have six colour genes; 6 in 64 will have five, 15 in 64 will have four, 20 in 64 will have three, 15 in 64 will have two, 6 in 64 will have one, and 1 in 64 will have no genes for the red colour.

Coat Colour in Man

The wheat seeds heterozygous for the R genes have only slightly less intense red colours than do the homozygotes, but Nilsson-Ehle noticed that the variation in the seed colour observed in the F_2 generation was greater than in the F_1. This is expected, since the F_1 consists of individuals that are genotypically uniform, whereas in the F_2 a segregation takes place. The hypothesis of multiple factors, proposed to explain the inheritance of quantitative characters, assumes that some character differences are determined by series of independent genes. The effects of these genes are *cumulative*, or *additive*; i.e., each gene contributes a certain quantity of colour, or height, or weight, or fertility, and the phenotype that develops shows a simple summation of these quantities. Dominance is usually absent or incomplete; the F_1 generation appears to have a blend of the characteristics of the parents. In the F_2 generation most individuals are still intermediate between the grandparents, and usually resemble the F_1 parents, as though the heredities of the strains crossed were thoroughly mixed or fused. But some F_2 individuals are intermediate between the F_1 and the grandparents, and finally, a minority of individuals may resemble the grandparents.

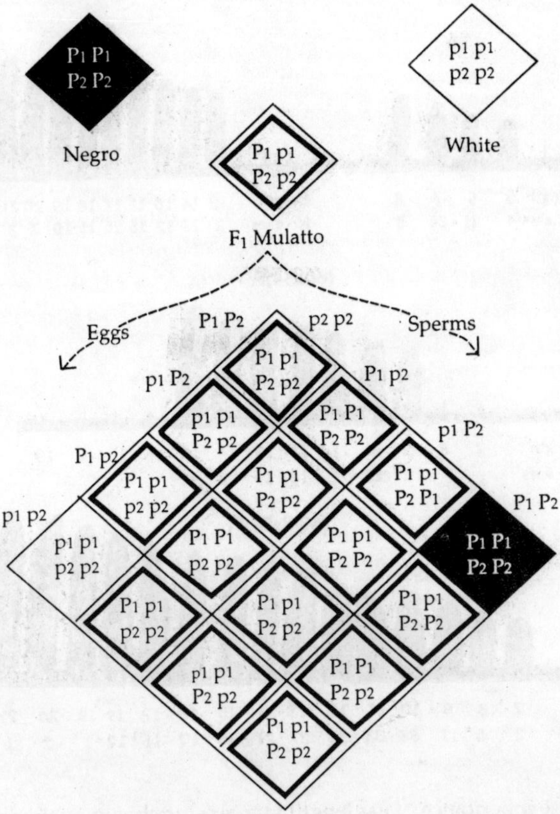

Fig. 15. The interaction of two hypothetical independent pairs of alleles (multiple factors) affecting skin colour, as in matings of Negroes and whites. The sixteen possible combinations of F_1 gametes with their genotypes and phenotypes are shown in the F_2 checkerboard.

Davenport (1913) saw that the hypothesis of multiple explains the inheritance of a character which seemed baffling and even incompatible with the gene theory, namely skin colour in mulattoes. He thought that the situation might be satisfactorily accounted for by assuming that Negroes native to central and western Africa differ from whites in two pairs of pigment-forming genes showing no dominance. Assume, then, that Negroes have four colour genes, $P_1P_1P_2P_2$, and whites no colour genes, $P_1P_1P_2P_2$. The first-generation mulattoes have two colour genes, P_1p_1 P_2p_2 and their colour is intermediate between the parents.

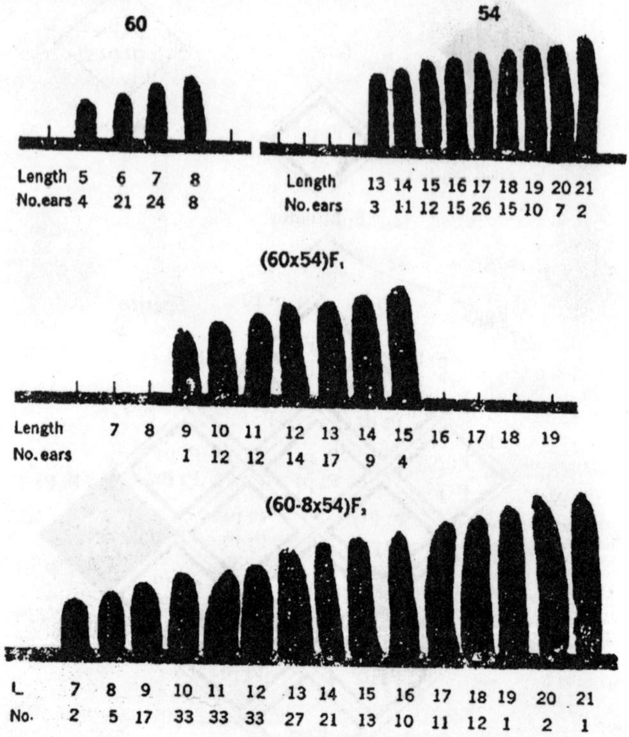

Fig. 16. The inheritance of ear length in maize, as shown by the results of crossing a short-eared variety of popcorn with a long-eared variety of sweet corn. Ears showing range in the length of the parent types are pictured above, with the F_1 and F_2 generations below. (*From East*)

The offspring of mulattoes (the F_2 generation of the Negro × white cross) would be like that shown in Figure for the colour of wheat kernels. More than a third (6 out of 16) of individuals among the F_2 mulattoes have two colour genes and hence resemble the F_1 mulattoes in skin pigmentation. But 4 out of 16 have three colour genes, and 4 out of 16 have only one. These people will be intermediate in skin pigmentation between the F_1 mulattoes and the original Negro parents, and between the mulattoes and the

Table 9
Inheritance of Quantitative Differences in Pied Spotting

	Per cent of dorsal white																				
	0	5	10	15	20	25	30	35	40	45	50	55	60	65	70	75	80	85	90	95	100
P_1 line 118 (F_{7-14})	..	7	89	35	2	1															
P_1 line 19 (F_{7-8})	..	..	..	..	..	..	..	..	..	..	..	..	..	..	..	..	..	1	9	29	134
F_1 line 19 × line 118	..	..	..	..	1	1	4	3	3	10	7	9	9	7	1	1					
F_2 line 19 × line 118	..	..	1	2	6	15	10	12	14	19	27	29	28	19	34	21	14	2	3	2	
BC F_1 × line 19	..	..	..	..	..	..	..	..	..	..	..	..	..	5	16	33	24	15	10	9	12
P_1 line 190*a* (F_{10-17})	..	..	3	10	22	23	31	27	28	15	5										
P_1 line 19 (F_{7-8})	..	..	..	..	..	..	..	..	..	..	..	..	..	..	..	..	..	1	9	29	134
F_1 line 190 *a* × line 19	..	..	..	..	..	..	..	..	..	..	..	..	1	5	4	1					
F_2 line 190 *a* × line 19	..	..	..	..	..	..	..	..	2	5	1	7	16	29	45	10	3	2	3	2	

original white parents respectively. Finally, 1 out of 16 F_2 individuals will have 4 colour genes, and a pigmentation as das dark as Negroes and 1 out of 16 will have no colour genes and a light skin resembling that of whites.

It is easy to predict also the skin pigmentation in the offspring of the backcrosses of the F_1 mulattoes ($P_1p_1P_2p_2$) with whites and with Negroes. In the backcross mulatto × white, half the offspring will have one pigment gene (either P_1 or P_2) and a skin colour intermediate between F_1 mulattoes and whites. One-quarter of the offspring will have two pigment genes and will resemble the F_1 mulattoes, and one-quarter will have no pigment genes and will resemble the whites. In the backcross mulatto × Negro, half the offspring will have three pigment genes, one-quarter two, and one-quarter four pigment genes. These people will range in skin pigmentation from that in F_1 mulattoes to that in Negroes. It is probable that Devenport's hypothesis oversimplifies the situation.

The difference in skin colour between Negroes and whites seems to be due to more than two pairs of pigment genes, possibly to four or even more pairs. After all, considerable variation in skin colour is observed among whites, and the variation is no less pronounced among African Negroes. Some of this variation assuredly has a genetic basis. These pigment genes are not necessarily equally potent; some may cause production of more pigment than others. But Davenport's hypothesis is essentially correct in that the heredity of skin pigmentation in man is due to the action of several colour genes with incomplete or with no dominance and with additive effects. Owing to gene segregation, some persons in the offspring of mulattoes resemble in skin colour, and also in other traits, people of "pure" white ancestry. One of the persistent popular beliefs is that such persons nevertheless carry Negro traits in their "blood" and that a black child may occasionally be born if they marry white persons.

It is easy to see that this belief conflicts with Davenport's hypothesis. A "throwback" or "atavism" is possible if both parties to a marriage are heterozygous carriers of same recessive gene. But the genes that govern skin pigmentation in man apparently show no dominance. If both parents carry such genes (for example, if their genotypes are $P_1p_1p_2p_2$ and $p_1p_1P_2P_2$ respectively), then some of their children can be either darker ($P_1p_1P_2p_2$) or lighter ($p_1p_1p_2p_2$) than either parent. But when one of the parents carries no pigment genes ($p_1p_1p_2p_2$), no individual in the offspring can be darker than the darker of the two parents.

Ear Size in Maize

One of the first successful applications of the polygene hypothesis was in East's study of the inheritance of the length and other dimensions of the ear in maize. He worked with two varieties: long-eared Black Mexican sweet corn and short-eared Tom Thumb popcorn. The ears have been arranged in classes differing by 1 cm in length, and under each class is given the number of individuals in that class. Thus, in the F_1 there was 1 plant in which the ear was 9 cm long, there was 12 in which it was 10 cm long, and so on. Each of the parental varieties varied somewhat in ear length, and the F_1 shows about the same degree of variability, although it is essentially intermediate between the two parents. In the F_2, however there is a much greater range, from plants with ears as short as the shorter-eared grandparent to those with ears as long as the longer-eared grandparent.

Table 10

Frequency Distributions Expected in F_2 Populations from Crosses of Parents Differing in a Dimensional Trait by 12 Units when the Difference in the Parents is Determined by One, Two, Three, and Six Pairs of Independent Genes, with Equal Additive Effects

Gene Pairs	*Class Centers* P_1						F_1						P_2	*n*
	101	102	103	104	105	106	107	108	109	110	111	112	113	
1	1,024	...	...	...	...	...	2,048	...	...	...	...	...	1,024	4,096
2	256	...	...	1,024	...	...	1,536	...	...	1,024	...	...	256	4,096
3	64	...	384	...	960	...	1280	...	960	...	384	...	64	4,096
6	1	12	66	220	495	792	924	792	495	220	66	12	1	4,096

There are relatively few of these extreme types and a relatively large number of those that have intermediate ear length essentially like that of the F_1. This situation is similar in its essentials to that of kernel colour in wheat, and although segregation into a series of distinct classes is impossible to demonstrate, the marked increase in variability of the F_2 as compared with the F_1 finds its simplest explanation in the segregation of a series of "multiple factors," each affecting ear length and all of them cumulative in their effect. A size difference between two strains of plants or animals may be due to one, two, or several gene differences.

A comparison of the kinds of segregations to be expected in the F_2 from parents whose differences in a quantitative trait such as size are due to 1, 2, 3 and 6 pairs of independent genes with equal and cumulative effects and without dominance. The frequencies will be recognised as proportional to the coefficients in the expansion of $(a + b)^n$ for the even powers from 2 to 12 corresponding to the number of alleles contributing to the total result. As the number of alleles increases, the relative frequency of the parental types recovered in the F_2 declines sharply and the different F_2 phenotypes are spread much more evenly over the range, approaching the type of continuous distribution obtained from the measurements of a continuously variable trait such as size in actual F_2 populations.

White Spotting at Mice

With polygenic inheritance, the number of genes segregating in the offspring of a cross may be hard to determine in the usual hybridization experiments. The reasons are that, first, the effects of individual genes on the phenotype may be obscured by variations due to environment, and, second, that the effects of one gene often interfere with the expression of other genes. In other words, the gene effects may not be simply additive, and gene interaction may complicate the situation. Among mice homozygous for the gene, *ss*, for pied spotting, animals occur ranging from those with only a small white spot on the belly to those with an entirely white coat.

By selection it is possible to produce strains of mice which differ in the range of spotted types they produce.

As in the cross of corn varieties differing in ear length, the F_1 is intermediate, whereas the F_2 is more variable, and a few individuals within the range of the parent types are recovered. In a backcross of the F_1 with animals of the nearly all-white stock, the parental type was recovered with about the frequency that would be expected if three or four genes with quantitative effects were segregating in the F_1. The results suggest that the lighter and darker spotted strains differ by several genes $s_1 \ldots s_n$, whose effects cumulate with each other and with those of *ss*, the pied gene being common to all these strains. This conclusion was later confirmed by separating the $s_1 \ldots s_n$ genes from *s* and analyzing them in outcrosses. Similar evidence has been obtained for the multiplicity of white-spotting factors in guinea pigs, in rats, and in rabbits.

Modifying Factors

The type of interaction in which several genes with small expression, such as s_2s_2, s_3s_3 above, exert their chief actions by changing the magnitude of effect of major gene, such as *ss* above, is sometimes referred to as modifying effect, and the minor genes, as modifiers. In this case the modifiers have a small effect even when the major gene is not present, so that the effects of the modifiers and of the major gene are qualitatively similar. An analysis of genes producing white spotting in guinea pig by Wright and Chase indicates a similar situation there, though probably with more genes concerned. Castle, by selecting lighter and darker variants of the hooded (white spotted) rat, produced strains that apparently differed by many modifiers with small effects, and this type of interaction is probably a common one. The complex of modifiers in a stock, which determines small effects and alters the expression of other genes introduced into it, is sometimes referred to collectively as the residual heredity of the stock. Because of the number of genes involved, it is usually difficult to analyze such complexes into single gene components.

Specific Modifiers

In a few cases it has been shown that multiple genes, or polygenes, exist which have little or no effect themselves on the particular trait under consideration, except in the presence of some specific main gene that they modify. Thus, in certain strains of mice, there occurs a form of spotting known as variegated. Such variegated mice are heterozygous for a dominant gene *W*, which is lethal when homozygous, *WW* mice usually dying shortly after birth with a severe anemia. When, rarely, they live long enough to develop hair, they are seen to be entirely white with black eyes; that is, the white spotting covers the entire coat.

Table 11
The Effects of Specific Modifiers *m (W)* on the Dominance of the Spotting Mutation *W* in House Mice

Parents	Progeny: Genotype for W and s	Condition of m(W) loci	Phenotype in per cent dorsal white	Proportions
Light variegated *WwSS* × *WwSS*	*WWSS*	$m_1m_2 \ldots m_x$	100	¾ white
	WwSS		90 – 95	
	wwSS		0 – 5	¼ unspotted
Unspotted	*WWSS*	$M_1M_2 \ldots M_x$	100	¼ white
WwSS × *WwSS*	*WwSS*		0	¾ unspotted
	wwSS		0	
Medium variegated	*WWSS*	$M_1m_2M_3m_x$	100	¼ white
WwSS × *WwSS*	*WwSS*		20–80	½ spotted
	wwSS		0	¼ unspotted
Black-eyed white	*WWss*	$m_1m_2 \ldots m_x$	100	¾ white
Wwss × *Wwss*	*Wwss*		100	
	wwss		50–90	¼ spotted
Intermediate spotted	*WWss*	$M_1M_2 \ldots M_3$	100	¼ white
Wwss × *Wwss*	*Wwss*		50 – 100	¼ white
	wwss		0 – 50	¾ spotted

The degree of white spotting of the heterozygotes, however, depends not only on *W* but on a complex of polygenes, each with a small effect, which determine the effect of *W*. In many of these modifiers are present, the *Ww* coat is nearly all white; if a few or none are present *Ww* mice have little or no white. These modifiers have, however, little or no effect in *ww* mice; even with the maximum number present, the animal is unspotted unless *W* is also present. One effect of these modifying genes is thus to determine the degree of expression or dominance of *W*. When all of the modifiers are present, the cross of *Ww* × *Ww* produces offspring in the ratio of ¾ white (*WW*, 2/4 *Ww*): ¼ unspotted (*ww*); when part of them are present, the offspring consist of ¼ white (*WW*), ½ variegated (*Ww*), and ¼ unspotted (*ww*); and when none is present, the progeny are ¼ white (*WW*) and ¾ unspotted (*Ww* and *ww*). This is a clear-cut demonstration of the fact that dominance in a pair of alleles is determined by the interaction of other genes.

Modifiers like these may be referred to as specific, since their expression seems to depend upon interaction with one specific genotype. This type of interaction probably intergrades imperceptibly with the more general, less specific modifying effects of other genes with quantitative effects. Because of the mutual interaction among many genes in producing a phenotype, it is probable that most genes may function as modifiers.

4

Linkage

When two or more than two genes are inherited together without any change for at least four to six generation, they are said to be the linked characters and the phenomenon is called linkage.

According to Sutton's hypothesis is correct, genes are parts of chromosomes. The inheritance of genes will therefore be identical with the inheritance of chromosomes. Stated in another way, all genetic theory must take into account the behaviour of chromosomes, and all chromosomes theory must take into account the results of genetic crosses.

The Prediction of Linkage

Sutton proposed his hypothesis, and pointed out one situation in which the Mendelian law could not apply, namely, those cases where two genes are carried on the same chromosome. Clearly, they would not obey Mendel's law of independent assortment. He foresaw that this problem would arise when more pairs of genes had been discovered than there are pairs of chromosomes in the species being studied. The cells of drosophile melaunogaster have four pairs of chromosomes as the diploid number. Let us assume that each of the first four mutant genes discovered are located on a different chromosome pair. If this is the case, each of these four genes will show independent assortment. If we neglect the case of those carried on the X, we shall obtain a ratio of 9 : 3 : 3 : 1 in the F_2 if the *P* generation cross was between an individual homozygous for both dominant genes and one recessive for both genes. What will happen when we discover the fifth pair of genes? Since there is no fifth pair of chromosomes, the fifth pair of genes must be located on a chromosome that already has one of the first four pairs of genes. When this situation arises, obviously the two pairs of alleles cannot act in an independent way during meiosis. They would be linked together in inheritance. This deduction is inevitable, if genes are on chromosomes.

Morgan stated that the pairs of genes of homozygous parent tend to enter in same gametes and to remain together, whereas same genes from heterozygous parents tend to enter in different gametes and remain apart from each other. He further stated that the tendency of linked genes remaining together in original combination is due to their location in same chromosome. The degree or strength of linkage depends upon the distance between the linked genes in the chromosome. Morgan's concept about the linkage developed the *theory of linear arrangement of genes* in the chromosomes which helped the cytogeneticists in the construction of genetic or linkage maps of chromosomes.

Chromosome Theory of Linkage

The chromosome theory of linkage of Morgan and Castle states that:

(i) The genes which show linkage, are situated in the same pair of chromosomes.

(ii) The linked genes remain arranged in a linear fashion on the chromosome. Each linked gene has a definite and constant order in its arrangement.

(iii) The distance between the linked genes determines the degree of strength of linkage. The closely located genes show strong linkage then the widely located genes which show weak linkage.

(iv) The linked genes remain in their original combination during the course of inheritance.

Sequence of Linked Genes

When two pairs of genes are linked, the linkage may be of either of two types in an individual heterozygous for both pairs (dihybrid): (1) the two dominant genes (e.g., R and Ro of sweet pea) may be located on one member of the chromosome pair; with the two recessive genes (e.g., r and ro) on the other, or (2) the dominant allele of one pair and the recessive of the other (R and ro) may be located on one chromosome of the pair, with the recessive of the first gene pair and the dominant of the second gene (e.g., r and Ro) on the other chromosome. The first arrangement, with two dominants on same chromosome is called cis arrangement. This type of interrelationship of dominant and recessive linked genes (e.g., RRo/rro) is called *coupling phase.*

The second arrangement, one dominant and one recessive gene on the same chromosome is called trans arrangements. Such type of interrelationship of dominant and recessived genes (e.g., Rro/rRo) is called *repulsion phase.*

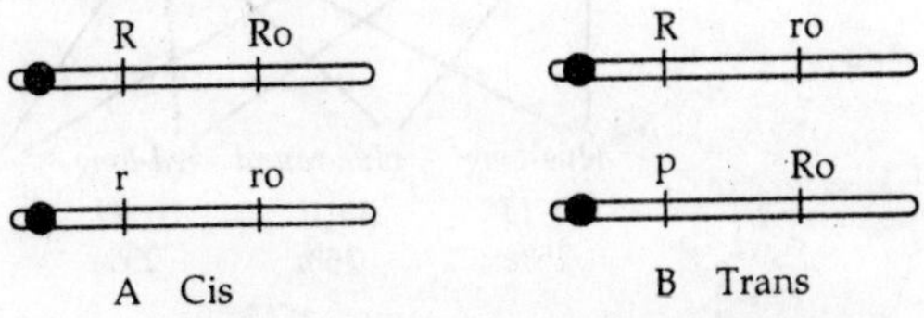

Fig. 1. Cis and trans arrangement of two pairs of linked genes of *Lathyrus odoratus.*

The Discovery of Linked Genes. In 1906 Bateson nd Punnett reported a cross involving two pairs of genes that did not show independent assortment. Their cross was with sweet peas, where *blue* flower colour (**B**) is dominant over *red* (**b**) and *long* pollen grain (**L**) is dominant over *round* pollen (**l**). The scheme of the cross was this:

P	phenotypic	*blue-long*	×	*red-round*
	genotype	BB LL		bb ll
	gametes	B L		b l
F_1	phenotypic	*blue-long*		
	genotype	Bb Ll		
	expected gametes	BL Bl bL bl		

If there is independent assortment in the Mendelian sense, we would expect the four classes of F_1 gametes shown in the diagram to be produced in equal numbers. Each type would account for 25 per cent of the total. The standard genetic way for finding gamete percentage is to cross the organism being tested with the pure recessive. This is called the *test cross.* For the F_1 heterozygous plant it would be this:

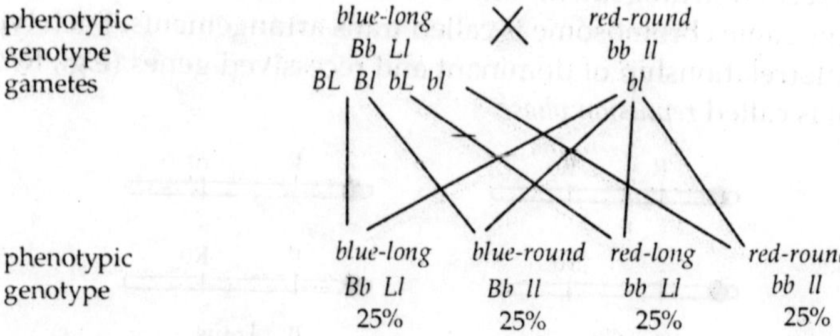

In this test cross, the phenotype of the offspring would be a measure of the gamete frequency of the plant being tested. Thus, 25 per cent of the gametes would be *BL* and 25 per cent of the offspring would be *blue-long*. This would be true for all classes of gametes, since they would be combining with a gamete having both recessive genes. When Bateson and Punnett made the cross, these were the results:

	Expected	*Actual*
blue-long	25 per cent	43.7 per cent
blue-round	25 per cent	6.3 per cent
red-long	25 per cent	6.3 per cent
red-round	25 per cent	43.7 per cent

Clearly these results do not conform to those expected from the Mendelian theory. Two points should be noticed.

1. The two most frequent phenotypes are those of the original parents (*blue long* and *red-round*).
2. The percentages of the original parental types (*blue-long* and *red-round*) are the same and the percentages of the two recombination classes (*blue-round* and *red-long*) are the same.

At even more disturbing finding was that the F_2 ratios depended largely on the genotype of the *P* generation. On strict Mendelian principles a cross of *blue-long* × *red-round* should give the same F_2 as *blue-round* × *red-long*. This was not so with Bateson's sweet peas. We have already seen that F_2 ratios were for the *blue-long* × *red-round* cross. The *blue-round* × *red-long* cross gave the following results:

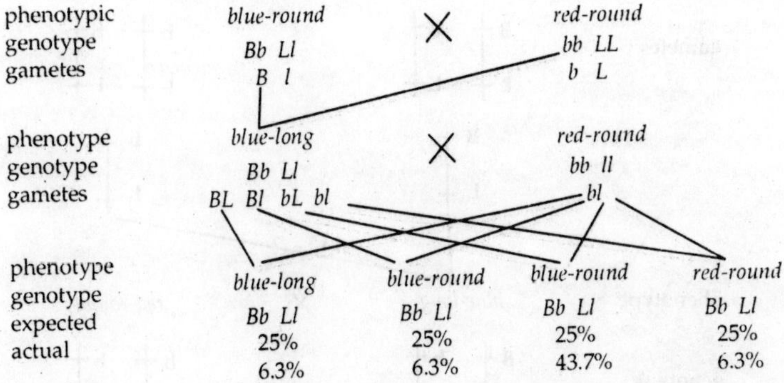

When these results are compared with those of the first cross, the percentages for each phenotype are found to be different, but again we notice a preponderance of the parental types. Both crosses suggest an orderly, though non-Mendelian, mechanism of inheritance. Some new principles must be involved. Let us try to explain the results on the basis of Sutton's hypothesis. If we assume that the two genes, *B* and *L* are parts of the same chromosome, this will be the schematic representation of the cross:

The percentages actually obtained can be compared with the ratios expected on the basis of Mendel's hypothesis and Sutton's hypothesis:

	blue-long	*blue-round*	*red-long*	*red-round*
Actual	43.7	6.3	6.3	43.7
Mendel's	25	25	25	25
Sutton's	50	0	0	50

At first sight it may seem as though Sutton's explanation is the better since the *blue-long* and *red-round* clusses are near the actual percentages. One difficulty, however, is fatal to Button's hypothesis. Both *blue-round* and *red-long* plants were obtained in the cross. Neither of these classes would be possible if both the *B* and *L* genes were part of the same chromosome. We may conclude that the results obtained in the cross cannot be understood on the basis of either Mendel's or Sutton's hypotheses, as these were originally stated.

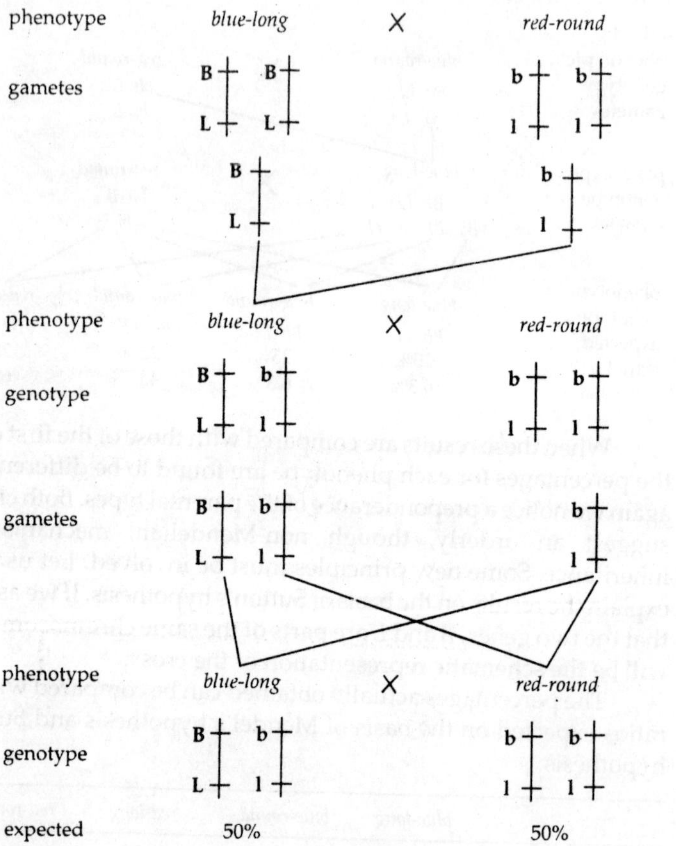

Coupling and Repulsion Hypothesis

W. Bateson, in 1905, described a cross in sweet pea, where a deviation from independent assortment was exhibited. The plants of a variety of sweet pea having blue flowers (B) and long pollen (L) were crossed with those of another variety having red flowers (b) and round pollen (l). The F_1 individuals (BbLl) had blue flowers and long pollen. These were crossed with plants having red flowers and round pollen (bbll). (In this case character for blue colour of flowers is dominant over red colour and long pollen character is dominant over round pollen. If independent assortment takes place, we should expect 1 : 1 : 1 : 1 ratio in a testcross. Instead 7 : 1 : 1 : 7 ratio was actually obtained, indicating that there was a tendency in the dominant alleles to remain together. Similar was the case with recessive alleles. This deviation was, therefore, explained as "gametic coupling" by Bateson.

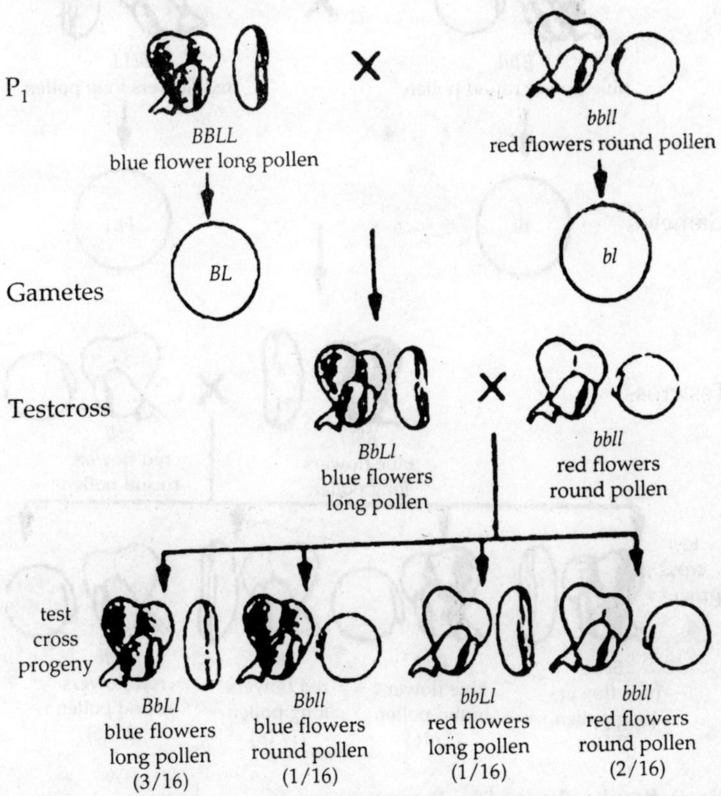

Fig. 2. Results obtained by Bateson in sweet pea when two characters, blue flowers and long pollens, were present in coupling phase.

Similarly, it was observed that when two such dominant alleles or two recessive alleles come from different parents, they tend to remain separate. This was called "repulsion". In Bateson's experiment in repulsion phase, one parent would have blue flowers and round pollen (BBll) and the other would have red flower and long pollen (bbLL). Bateson explained the lack of independent assortment in the above experiments by means of a hypothesis known as "coupling and repulsion hypothesis". Although coupling and repulsion was explained above were later discovered to be

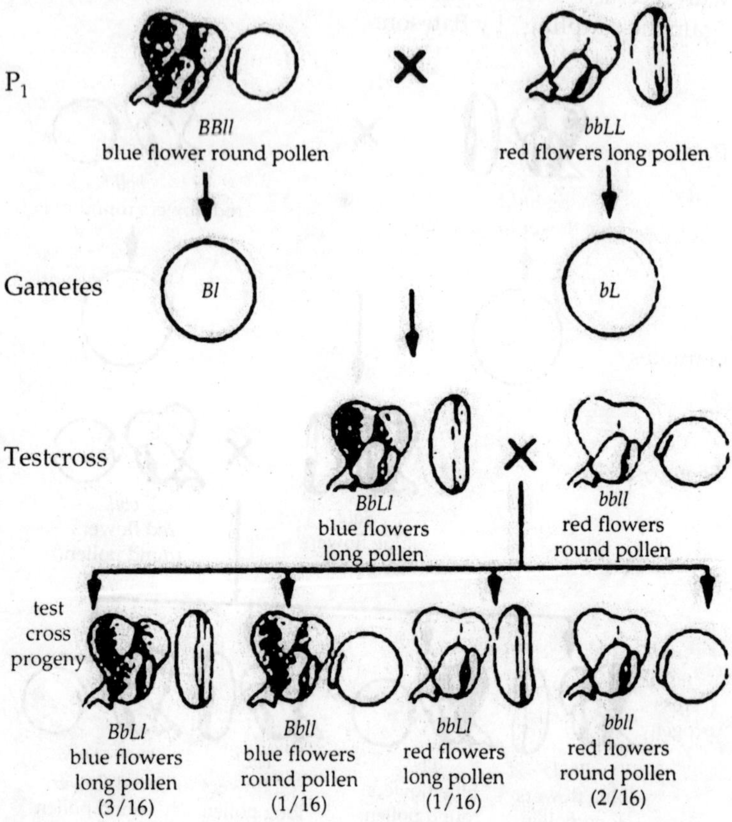

Fig. 3. Results obtained by Bateson in sweet pea when two characters, blue flowers and long pollen, were present in repulsion phase.

the two aspects of the same phenomenon called "linkage", the terms "coupling phase" and "repulsion phase" are still considered to be useful terms in scientific literature.

Morgan, while performing experiments with *Drosophila,* in 1910 found that coupling or repulsion was not complete. He proposed that the two genes are found in coupling phase or in repulsion phase, because these are present on the same chromosome (coupling) or on two different homologous chromosomes (repulsion). Such genes are then called as linked and the phenomenon is called linkage. He further suggested that the strength of linkage will be determined by the distance between the two genes in question. The greater this distance, lower will be the linkage strength. The linkage is broken down due to the phenomenon of crossing over occurring during meiosis.

Fig. 4. Exchange of chromosome segments during crossing over. The first figure represents a pair of normal chromosomes, the second figure represents exchange due to double crossing over.

Crossing over will be relatively more frequent, if the distance between the two genes is more than in case where the distance between two genes is lesser. The phenomenon of crossing over involves exchange of chromosome segments.

In order to make the phenomenon of linkage and crossing over easily understandable, let us take another hypothetical example, where genes A and B are involved, a and b being their recessive alleles respectively. A cross AB/AB × ab/ab would give rise to a F_1 dihybrid Ab/ab. This dihybrid will then be crossed with double recessive parent ab/ab to get the testcross progeny.

Depending up the distance between any genes which is inversely proportional to the strength of linkage, the non-crossovers will vary from 50 per cent–100 per cent (100 per cent non-crossovers is a state where no crossing over takes place as in male *Drosophila*. The crossovers will similarly vary from 0–50 per cent and will never exceed 50 per cent. The easiest way of finding out the proportion of non-crossovers and crossovers is to make a testcross (AB/ab × ab/ab) as shown in Fig. 4.4. In such a situation, since the phenotypic ratio will depend upon the ratio of different kinds of gametes coming from F_1 (AB/ab), the relative proportion of non-crossovers and crossovers can be easily determined from the phenotypic ratio.

Kinds of Linkage

The phenomenon of linkage is of following two kinds:

1. Complete linkage. When the linked genes are so closely located in chromosomes that they inherit in same linkage groups for two or more generations in a continuous and regular fashion, then, they are called *completely linked genes* and the phenomenon of inheritance of completely linked genes is called *complete linkage.*

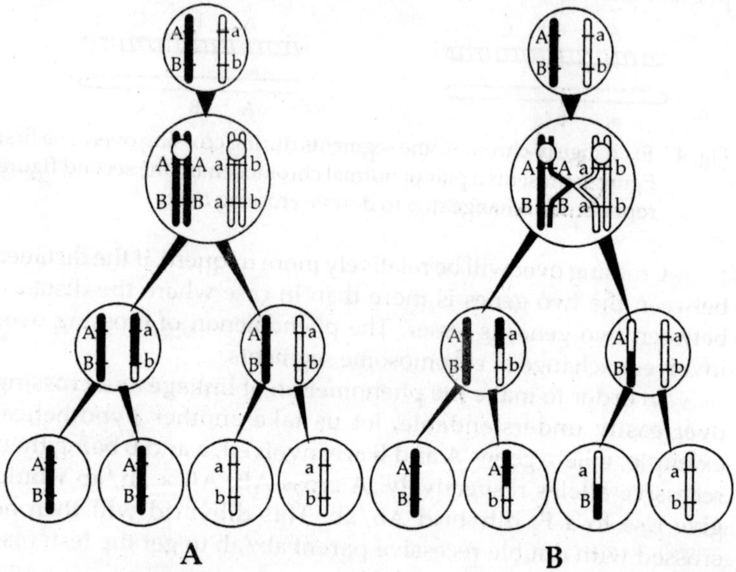

Fig. 5. A—Diagram of the segregation of two pairs of allelomorphic genes localised on the same pair of chromosomes without crossing over. B—Diagram of the segregation of two pairs of allelomorphic genes on the same chromosome between which crossing over takes place during meiosis.

Example: According to Bridge all the genes of male *Drosophila* remain completely linked. Further, in a mutant strain of *Drosophila,* the genes for bent wings (b^+) and shaven bristles (svn) of the fourth chromosome exhibit complete linkage.

2. ***Incomplete linkage.*** The linked genes do not always stay together because homologous non-sister chromatids may exchange segments of varying length (which bearing many linked genes) with one another during meiotic prophase, by

the process of crossing over. The linked genes which are widely located in chromosomes and have chances of separation by crossing over are called *incompletely linked genes* and the phenomenon of their inheritance is called *incomplete linkage.*

Examples: Incomplete linkage has been observed in pea, *Zea mays* (maize), tomato, female *Drosophila,* mice, poultry and man. Here, the examples of linkage have been considered only for *Drosophila* and *Zea mays* (maize).

Drosophila

The wild type Drosophila has gray body and long wings (b^+v^+/b^+v^+), where alleles for gray b^+ and long v^+ dominate over the mutant alleles for black b and vestigial v. When, a gray long fly (b^+v^+/b^+v^+) is crossed with a black bodied and vestigial winged fly (bv/bv), the F_1 heterozygote is found with gray long phenotype and b^+v^+/bv genotype. The F_1 heterozygote (b^+v^+/bv) when test crossed with double recessive parent (bv/bv), instead of occurring of two class of phenotypes in the ratio of 1 : 1, occur four classes of phenotypes. The test cross results are clearly showing that parental combinations (gray long and black vestigial) are those expected from complete linkage and appeared in 83 per cent cases. The other two (gray vestigial and black, long) are new combinations and appeared in 17 per cent cases. Thus, in 17 per cent cases crossing over has occurred.

Maize

Hutchison observed a case of incomplete linkage between the alleles for colour and shape of the seed in *zea mays* (maize). When a maize plant with seeds having colour and full endosperm (CS/CS) is crossed with another plants having recessive alleles for colourless, shrunken seeds (cs/cs), the F_1 heterozygotes are found with the phenotype of coloured full and genotype of CS/cs. When F_1 hybrid is test crossed with double recessive parent (cs/cs) four classes of descendants are obtained instead of two as shown in following figure:

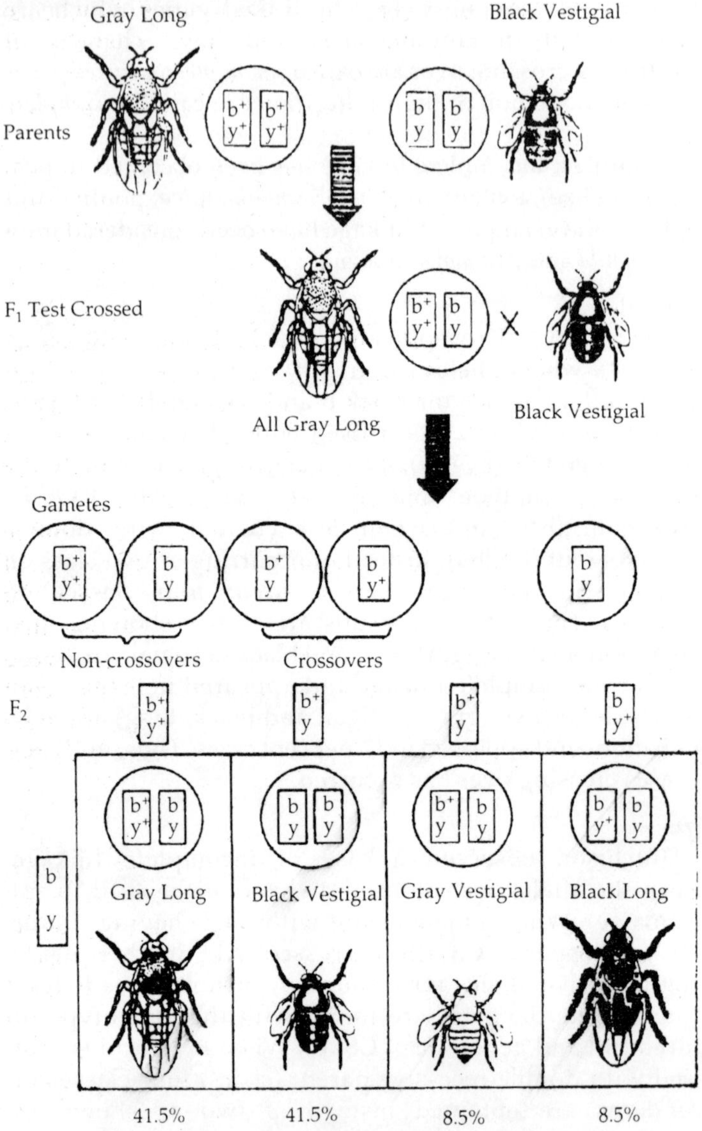

Fig. 6. Diagram of a cross involving linkage and crossing over. The genes for vestigial versus normal (long) and black versus gray body in *Drosophila* are linked, they are located in the same chromosome.

Parents:	Coloured full CS/CS	×	Colourless shrunken cs/cs	
$\mathbf{F_1}$		Coloured full (CS/cs)		
Test cross:	F_1 Coloured full CS/cs	×	Colourless shrunken cs/cs	

Test cross results:

Coloured, full	Coloured, shrunken	Colourless, full	Colourless shrunken
CS/CS	Cs/cs	cS/cs	cs/cs
48 per cent	2 per cent	2 per cent	48 per cent

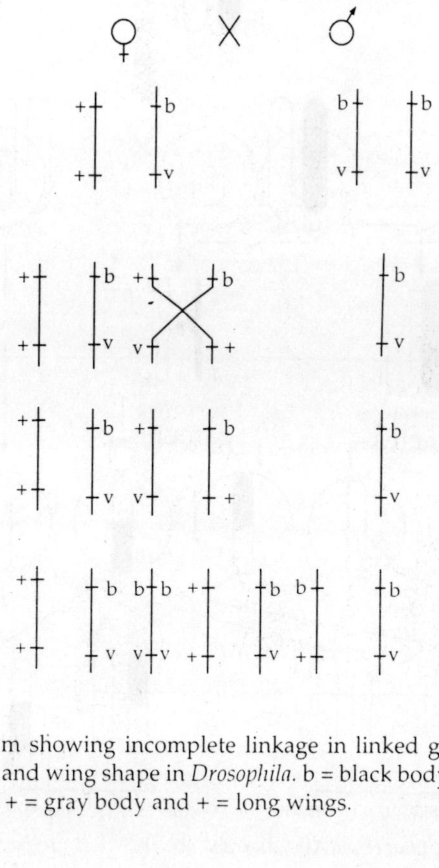

Fig. 7. Diagram showing incomplete linkage in linked genes for body colour and wing shape in *Drosophila.* b = black body, v = vestigial wings, + = gray body and + = long wings.

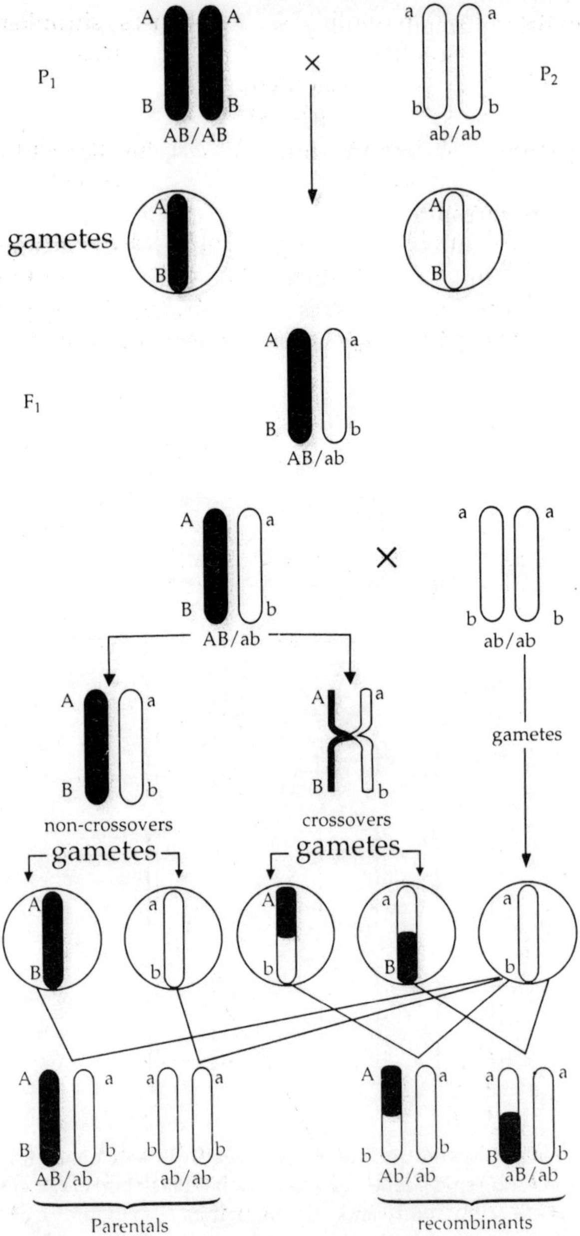

Fig. 8. A testcross (AB/ab × ab/ab) showing crossover and noncrossover gametes (from AB/ab) and progenies.

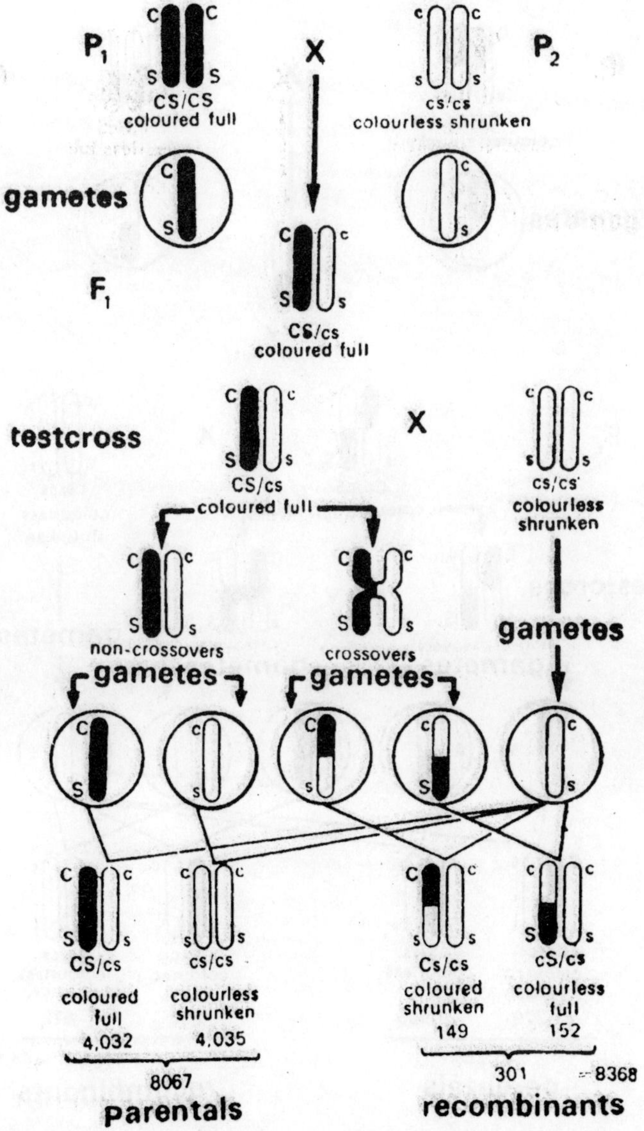

Fig. 9. Results obtained in maize by Hutchinson, when characters coloured and full seeds were taken in coupling phase.

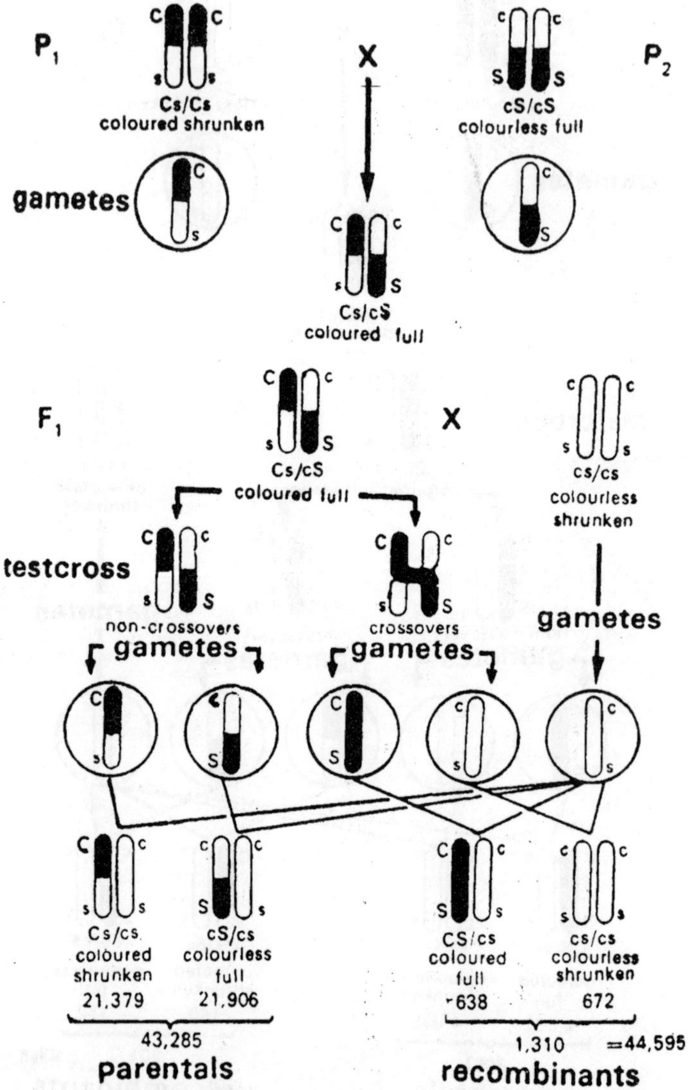

Fig. 10. Results obtained by Hutchinson in maize, when characters coloured seeds and full seeds were taken in repulsion phase.

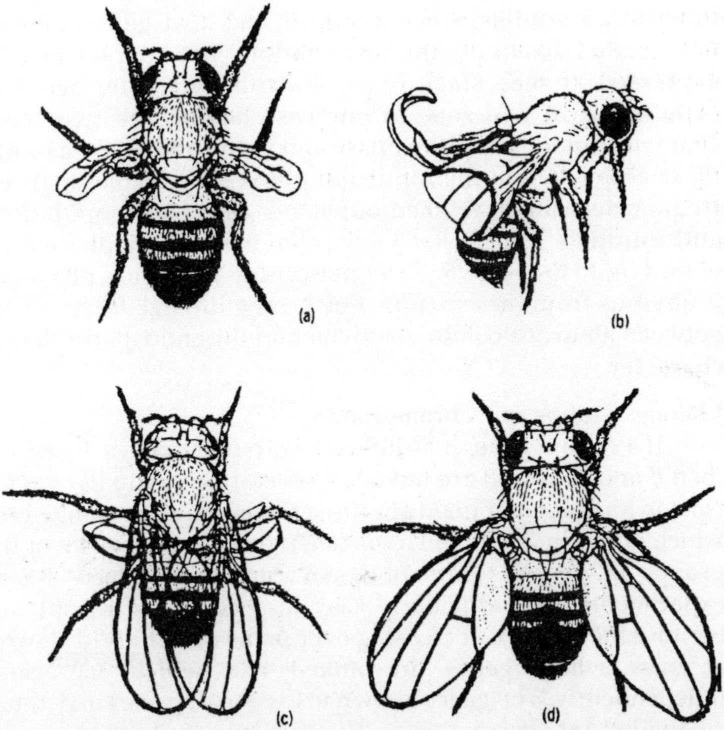

Fig. 11. Mutants of Drosophila melanogaster: (a) vestigial, (b) culry wings, (c) bithorax, (d) dichete.

The test cross results are clearly showing that parental combination of alleles (e.g., CS/CS and cs/cs) are those expected from complete linkage and appear in 96 per cent cases, the other two are new combinations (*e.g.*, Cs/cs and cS/cs) and appear in 4 per cent cases. Thus, in 4 per cent cases crossing over have occurred between linked genes.

A Testcross in Maize

Hutchinson made use of two characters known to be located on chromosome IX in maize. These two characters were coloured aleurone (C) vs colourless aleurone (c) and full endosperm (Sh) vs shrunken endosperm (sh) in the seed of maize crop. The seed characters have a special utility in inheritance studies, since a crop in the next generation is not needed to study the segregation. The characters are expressed at seed stage itself. Hutchinson performed the experiment in two sets, in one case having the two seed characters in the coupling phase and in the other case having these characters in the repulsion phase. The recombination frequencies can be worked out as per cent total population and would be 301/8368 = 3.6 per cent in coupling phase and would be 1310/44, 595 = 2.94 per cent in repulsion phase. It is obvious from these results that strong linkage force exists between aleurone colour character and the endosperm shape character.

Linkage Groups and Chromosomes

If a certain gene, *A*, is linked to two other genes, *B* and *C*, then *B* and *C* as well are linked. If several genes are known to exist in an animal or plant species, crosses may be arranged in which either independent assortment or linkage of pairs or of groups of several pairs of genes can be observed. Such experiments are particularly easy in species of Drosophila, because the absence of crossing over in the male hybrid shows at once which genes are linked and which segregate independently. The genes known to exist in a species may thus be divided into *linkage groups*. The members of a linkage group show linkage to each other. In genetically well-studied species, the number of linkage groups is equal to the number of chromosome pairs which the species possesses. Several species of Drosophila have been widely used for studying linkage. Most of the early work was done on *D. melanogaster*, a vinegar fly, sometimes incorrectly referred to as fruit fly. These flies are very easy to breed under laboratory conditions; their culture is simple and inexpensive; their life cycle is less than two weeks long at 75°F; their fecundity is so high that a pair of parents may produce hundreds and even thousands of offspring; they produce numerous hereditary variants (mutations); and their chromosomes, especially in the cells of larval salivary glands, are uncommonly favourable for observation.

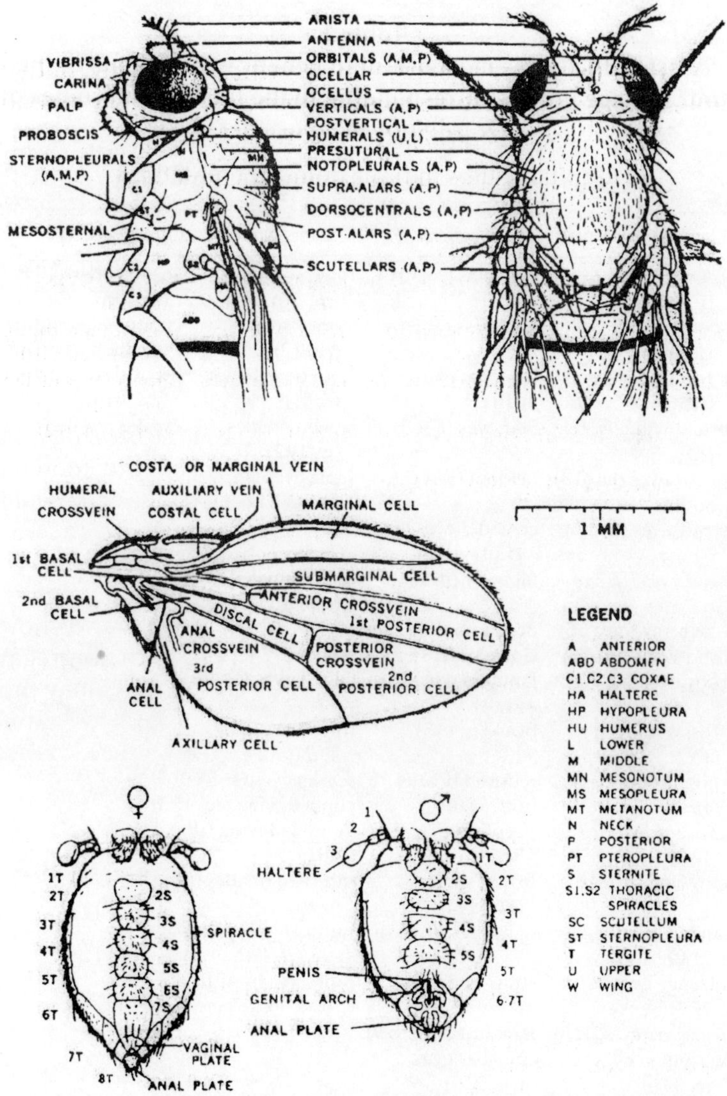

Fig. 12. Details of the external structure of *Drosophila melanogaster*.

Table 1
A list of the genes most useful in genetic experiments in the four linkage groups corresponding to the four chromosomes of *Drosophila melanogaster*

Capital Letters indicate dominant mutations

Chromosome X or 1	*Chromosome II*	*Chromosome III*	*Chromosome IV*
yellow body (*y*, 0.0)	net veins (*net*, 0.0)	roughoid eyes (*ru*, 0.0)	shaven bristles (*sv*, 0.0)
scute bristles (*sc*, 0.0)	aristaless (*al*, 0.0)	veinlet wing (*ve*, 0.2)	cubitus-interruptus venation (*ci*, 0.0)
silver body (*svr*, 0.0)	ochracea eyes (*ocr*, 0.0)	Roughened eyes (*R*, 1.4)	grooveless scutellum (*gvl*, 0.0)
broad wing (*br*, 0.6)	Star eyes (*S*, 1.3)	javelin bristles (*jv*, 19.2)	eyeless (*ey*, 0.2)
prune eyes (*pn*, 0.8)	heldout (*ho,4.0*)	sepia eyes (*se*, 26.0)	
white eyes (*w*, 1.5)	echinoid eyes (*ed*, 11.0)	hairy body (*h*, 26.5)	
facet eyes (*fa*, 3.0)	fat body (*ft*, 12.0)	approximated veins (*app*, 37.5)	
echinus eyes (e.c, 5.5)	dumpy wing (dp, 13.0)		
bifid veins (bl, 6.9)	clot eyes (cl 16.5)		
ruby eyes (*rb*, 7.5)	Jammed wing (*J*, 41.0)	Glued eyes (*Gl*, 41.4)	
crossveinless (*cv*, 13.7)	black (*b*, 48.5)	thread arista (*th*, 43.2)	
roughex eyes (*rux*, 15.0)	reduced bristles (*rd*, 51.0)	scarlet eyes (*st*, 44.0)	
		clipped wing (cp. 45.3)	
carmine eyes (*cm*, 18.9)	purple eyes (*pr*, 54.5)	Wrinkled wing (*W*, 46.0)	
cut wing (*ct*, 20.0	Bristle short (*Bl*, 54.8)	inturned bristles (*in*, 47.0)	
singed bristles (*sn*, 21.0)	light eyes (*lt*, 55.0)	radius-incompletus veins (*ri*, 47.1)	
lozenge eyes (*lz*, 27.7)	straw body (*stw*,	pink eyes (*p*, 48.0)	
respebrry eyes (32.8)	tarsi fused (ti, 55.9)		
vermilion eyes (*v*, 33.0)	cinnabar eyes (*cn*, 57.5)		
miniature wing (*m*, 36.1)	engrailed (*en*, 62.0)		
dusky wing (*dy*, 36.2)	scabrous eyes (*sca*, 66.7)	blistery wing (*by*, 48.7)	
sable body (*s*, 43.0)	vestigial wing (*vg*, 67.0)	mussed wing (*mu*, 50.0)	

garnet eye (*g*, 44.4)	Lobe eye (*L*, 72.0)	karmoisin eyes (*kar*, 52.0)
scalloped wing (*sd*. 51.5)	curved wing (*c*, 75.5)	Stubble bristles, (*Sb*, 58.2)
uneven eyes (*un*, 54.4)	plexus veins (*px*, 100.5)	spineless bristles (*ss*, 58.5)
forked bristles (*f*, 56.7)	brown eyes (*bw*, 104.5)	Roof wing (*Rf*, 59.0)
Bar eyes (*B*, 57.0)		Stripe thorax (*sr*, 62.0)
Beadex wing (*Bx*, 59.4)		glass eyes (*gl*, 63.1)
fused veins (*fu*, 59.5)		Delta veins (*Dl*, 66.2)
bobbed bristles (*bb*, 66.0)		Hairless bristles (*H*, 69.5)
		ebony body (*e*, 70.7)
		Prickly bristles (*Pr*, 90.0)
		rough eyes (*ro*, 91.1)
		Beaded wing (*Bd*, 93.8)
		claret eyes (*ca*, 100.7)
		brevis bristles (*bv*, 104.3)

Many mutations have been observed in *D. melanogastern* since about 1909, when this species was first used for such studies by T.H. Morgan and his collaborators. The numbers of main gene loci belonging to the four linkage groups in *D. melanogaster* are, according to Bridges and Brehme (1944), as follows: first 141; second, 228; third, 156; fourth, 12. The species possesses, as we know, four pairs of chromosomes (in other words, the haploid chromosome number is n = 4). There are two rod-shaped X chromosomes in females and only one in males, the partner of that in the male being the hook-shaped Y chromosome. Since the genes of the first linkage group all show the sex-linked inheritance, they evidently are localised in the X chromosomes. The fourth linkage group contains the fewest genes; this makes it probable that this group corresponds to the small dotlike pair of chromosomes. The second and the third linkage groups correspond to the two large V-shaped chromosome pairs, called the second and third chromosomes.

In species of *Drosophila* other than *D. melanogaster*, the haploid chromosome numbers vary from 3 to 6 and the numbers of linkage groups vary accordingly. In other genetically well-studied organisms, a similar correspondence is observed. Thus, maize has 10 pairs of chromosomes, and approximately 400 genes studied in this plant fall into 10 linkage groups, whereas in peas. *Pisum sativum*, there are 7 linkage groups and 7 pairs of chromosomes. As could be expected, in several organisms the number of linkage groups known is less but is nowhere greater than the haploid number of chromosomes. Thus, in the mouse 16 linkage groups are known for 20 pairs of chromosomes, in the rabbit 11 linkage groups for 22 pairs of chromosomes, in the tomato 10 linkage groups for 12 pairs of chromosomes, etc.

Factors Affecting the Frequency of Linkage

Frequency of chiasma. The frequency of crossing over between genes, and hence formation between their loci in the chromosome, may be influenced by a number of physiological and environmental factors. Thus, with the increasing age of the female of *D. melanogaster*, the amount of crossing over in her eggs becomes smaller. Furthermore, this effect is not equally pronounced in all chromosomes but is particularly noticeable in the middle portion of the second and third chromosomes in the neighbourhood of the loci of the centromeres. Temperature, X rays, and the chemical composition of the food are other agencies that modify the frequency of crossing over.

The frequency of crossing over in *Drosophila* males is so low as to be negligible under normal circumstances, but Whitting-hill has shown that it can be increased by X-rays. Although crossing over normally occurs only in gametogenesis or sporogenesis during meiosis, under some exceptional circumstances not yet understood, it may take place in somatic cells. This *somatic crossing over* has been studied by Stern in Drosophila and by Jones in maize. It results in the appearance of "twin spots" of tissue somewhere in the body which show complementary genetic constitutions. Suppose, for example, that a Drosophila female carries in one of her X chromosomes the gene for singed bristles. *sn* and in her other X that for yellow body, *y*. Such a female has, of course, normal bristles and a gray body, because the genes *sn* and *y* are recessive to their normal alleles. But if a cell of this constitution, + *sn*/*y* +, undergoes a meiosis-like process, cells homozygous for *y* and others homozygous for *sn* will be produced. The resulting fly will be in general normal but will have on some part of its body a yellow spot with normal bristles and next to it a spot with singed bristles of normal colour.

Significance of Linkage

The phenomenon of linkage has one of the great significance for the living organisms that it reduces the possibility of variability in gametes unless crossing over occurs.

5

Crossing Over

During gametogenesis the homologous chromosomes exchange their segments meiosis. During prophase I of meiosis, the homologous chromosomes are brought together during the process of synapsis at zygotene. Each chromosome also divides longitudinally into two chromatids at this stage. At pachytene, the chromatids exchange segments. The homologous centromeres then start separating away from one another (terminalisation) and figures of chiasmata can be observed. It should be realised that chiasmata are the result and not the cause of crossing over.

The situation was eventually clarified by Morgan. New mutants were discovered in considerable numbers in his Drosophila experiments. He found that many crosses involving two pairs of genes gave the independent assortment expected in the Mendelian scheme. In other crosses, deviations of the sort found by Bateson and Punnett were encountered. Morgan was convinced of the correctness of Sutton's hypothesis that genes are parts of the chromosomes. He believed that the exceptions to independent assortment were due to the two different genes being on the same chromosome. But to explain the results satisfactorily, it was necessary to assume that under some circumstances genes could be transferred from one chromosome to another. A possible cytological basis for an exchange of genes was provided by Janssens. He described a type of behaviour of chromosomes in meiosis that is now known to be of nearly universal occurrence in both animals and plants. It is called *crossing over.*

Crossing over occurs during the tetrad stage. Synapsis results in homologous chromosomes coming close together with their long axes parallel. Both chromosomes duplicate and a tetrad of four chromatids is formed. According to Janssens, there is considerable coiling of chromatids around one another at this time. In some cases two of the chromatids break at the corresponding place on each. The broken chromatids may rejoin in such a way that a section of one chromatid is now joined with a section of the other. As a result, 'new' chromatids are produced that are mosaics of segments of the original ones.

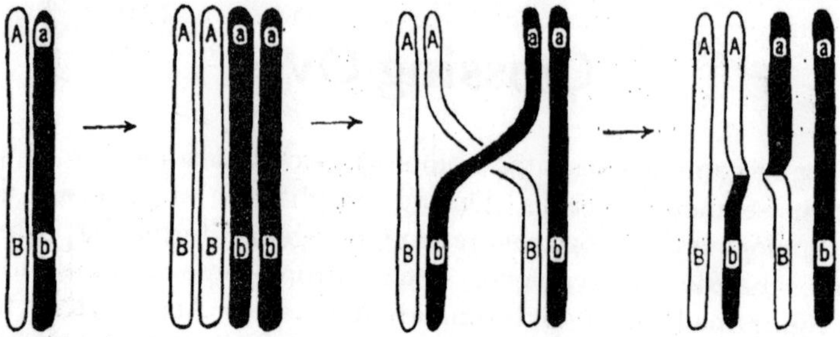

Synapsis of homologous chromosomes — Duplication of chromosomes to form tetrads — Two chromatids bend across one another — The chromatids break at the point of contract and fuses with portions of the other chromatid

Fig. 1. Janssens' theory of crossing over.

Janssens' theory of crossing over could provide the basis of gene transfer from one chromosome to another, and Morgan suspected that it did. This is what Morgan wrote:

In consequence, the original materials will, for short distances, be more likely to fall on the same side of the split, while remoter regions will be as likely to fall on the same side as the last, as on the opposite side. In consequence, we find linkage in certain characters, and little or no evidence at all of linkage in other characters; the difference depending on the linear distance apart of the chromosomal materials that represent the factors. Such an explanation will account for all of the many phenomena that I have observed and will explain equally, I think, the other cases so far described. The results are a simple mechanical result of the location of the materials in the chromosomes, and of the method of union of homologous chromosomes, and the proportions that result are not so much the expression of a numerical system as of the relative location of the factors in the chromosomes. Instead of random segregation in Mendel's sense we find 'association of factors' that are located near together in the chromosomes. Cytology furnishes the mechanism that the experimental evidence demands. *Crossing over* was the term applying to the coiling, breaking, and rejoining of homologous chromosomes during meiosis.

The following is an example of inheritance of two linked genes: In Drosophila *gray* body colour (B) is dominant to *black* body colour (b). *Long* wing (V) is dominant to *vestigial* wing (v). The two pairs of genes are located on the same pair of autosomes.

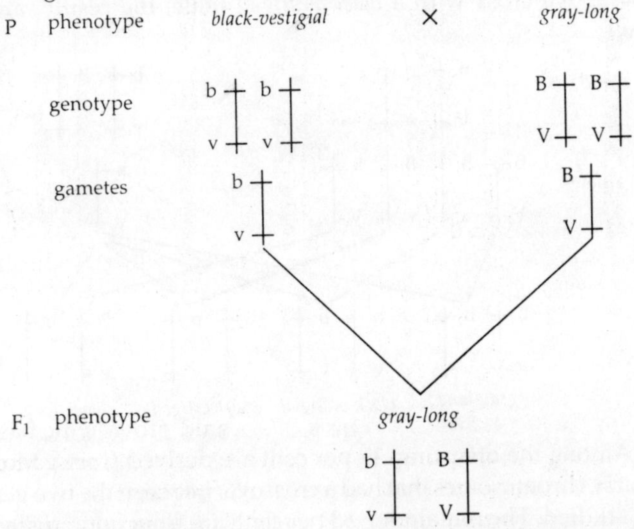

This F_1 individual will produce four types of gametes, two the result of crossing over and two non-crossovers. Since crossing over occurs in the tetrad stage we could diagram gamete formation as follows:

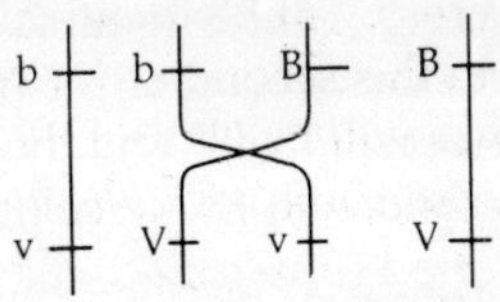

The middle pair of chromatids will break and recombine. The two meiotic divisions then occur and each resulting gamete will receive one chromosome. The four types of gametes will be as follows:

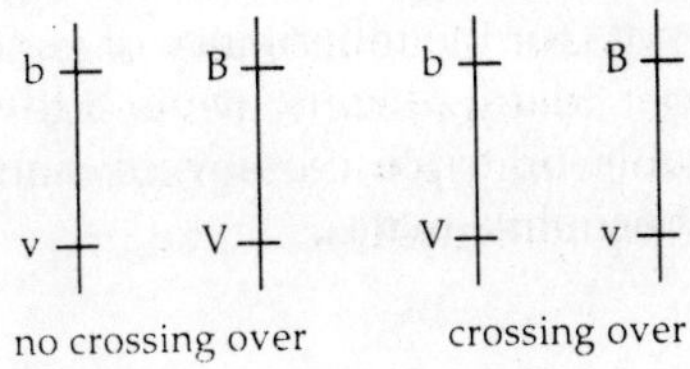

These four gametes are not produced in equal frequency. Morgan pointed out that *the chance of a crossover occurring between two genes is a function of the distance between them.* The closer they are the smaller the chance of crossing over. If we mate an F_1 female of the above cross with a *black-vestigial* male, the results are as follows:

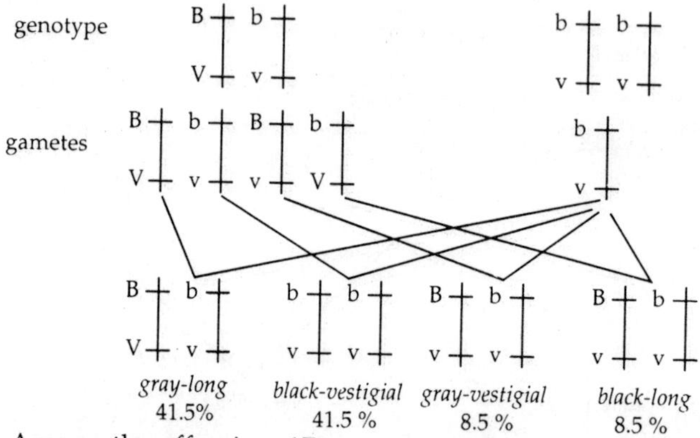

Among the offspring, 17 per cent are derived from gametes that carry chromosomes that had a crossover between the two genes being studied. The remainder, 83 per cent, are from non-crossover gametes.

The frequency of crossing over between any two genes is nearly constant. If we cross a *black-long* × *gray-vestigial* fly the F_1 would be *gray-long* and heterozygous for both genes. In these two respects the F_1 of this cross will be identical with that in the one previously described. In this second cross, however, one of the P generation chromosomes will be *bV* and the other *Bv*. If a female of this constitution is crossed with a *black-vestigial* male the offspring will be as follows:

41.5 per cent *black-long*
41.5 per cent *gray-vestigial*
8.5 per cent *gray-long*
8.5 per cent *black-vestigial*

Subsequent events have shown that Morgan's explanation satisfactorily accounts for the inheritance of genes located on the same chromosome. Many details were added, such as the occurrence of double or triple crossovers, and the absence of crossing over in Drosophila males.

The analysis of these experiments is an example of the mutual checking that the combined genetic and cytological approach permits. As we have mentioned before, inheritance must be explained in both fields. If two genes are in the same chromosome they will be linked in inheritance. If they are not completely linked there must be a chromosomal basis for the recombination. A chromosomal basis is to be found in the phenomenon of crossing over.

According to its occurrence in the germinal or somatic cells following two types of crossing over have been recognised.

Germinal or meiotic crossing over. Commonly crossing over occurs only in the germinal cells of reproductive organs during the process of gametogenesis which includes meiosis. This type of crossing over is called *germinal* or *meiotic crossing over*. It is universal in its occurrence and has great genetic significance.

Somatic or mitotic crossing over. Sometimes crossing over may occur during mitosis of somatic cells. This type of crossing over occurs in rare cases, has no genetic significance and is called *somatic* or *mitotic crossing over*. It has been observed in body cells of *Drosophila* by *stern* and in the fungus *Aspergillus nidulans* by *Pontecorvo*.

In somatic crossing over two chromosomes with unequal chromatids are formed. In the following division, in 50 per cent of cases two different alleles go to the same daughter nucleus and the original situation is restored. In the other 50 per cent, identical alleles go to the same pole, a situation comparable to anaphase II segregation with double reduction. Thus, in one cell two recessive alleles are present and recognisable, marking the event of crossing over.

Complex contains DNA and a specific proteinaceous material called synaptinemal complex material.

The synaptinemal complex serves crossing over by facilitating effective synapsis. Robert King (1970) suggested that synaptinemal complex may orient the non-sister chromatids of homologs in a manner to facilate it enzymatically induced exchanges between their DNA molecules. Comings and Okada (1971) have shown electron microscopically that synapsis occurs at two levels—one at chromosomal level and the other at molecular level. According to them, the synaptinemal complex pulls homologous chromosomes in the approximate association with each other but plays no role in molecular pairing of DNA strand.

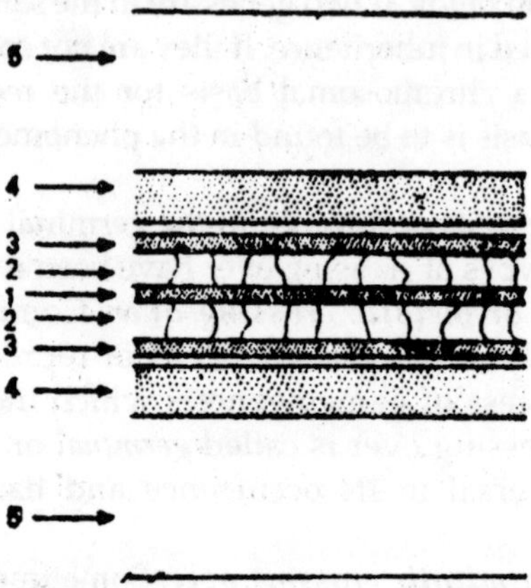

Fig. 2. The components of the synaptinemal complex.

One important fact about crossing over that has been established by cytological observation and then confirmed by breeding experiments is that the exchanges of chromosome segments which cause recombination of linked genes occur at the four-strand, or tetrad, stage, in the prophase of meiosis. At this time, the homologous chromosomes have paired, each homologue has split into two chromatids held together by the undivided centromere, and chiasmata are observed in the tetrad. A careful examination shows that at each chiasma only two of the four chromatids cross over, whereas the two other chromatids preserve their original continuities. In genetic terms, this must mean that of the four chromosomes that emerge from meiosis, only two can show crossing over at any one level. In most animals and plants it is impossible to determine by observing the gene recombination in linkage experiments whether the crossing over that produced this recombination involved two, three, or all four strands of chromatids.

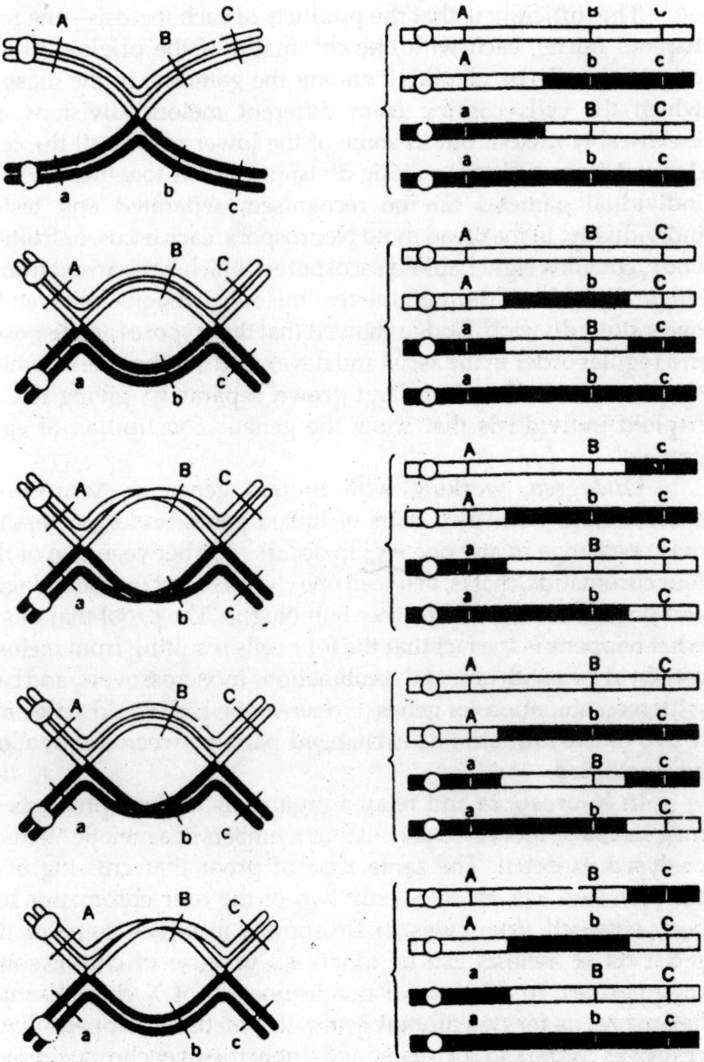

Fig. 3. Diagram showing chromosome bivalents with single chiasmata (above) and double chiasmata (below). 1. Single chiasma. 2. Two-strand double crossover. 3. Four-strand double crossover. 4 and 5. Three-strand double crossover.

The difficulty is that the products of each meiosis—the four haploid nuclei, each with one chromatid of the original four—cannot usually be identified among the gametes, in the mass of which the cells coming from different meiotic divisions are inextricably mixed. But in some of the lower plants, all the cells derived from a single meiotic division remain together and the individual gametes can be recognised, separated and tested individually. In the bread mold Neurospora, each ascus, or fruiting body, contains eight haploid ascospores, which have arisen from a single diploid cell through the two meiotic divisions followed by one mitotic division. Dodge showed that these spores are disposed in a regular order in the ascus and developed a technique by which they can be dissected out and grown separately, giving rise to haploid individuals that show the genetic constitution of each gamete.

Lindegren, working with mutant genes in Neurospora showed that, when two pairs of linked genes undergo crossing over, exchange in any one region occurs only between two of the four chromatids, that is, between one chromatid of one homologue and one chromatid of the other homologue. The proof that this is what happens is the fact that the four cells resulting from meiosis consist of two with parental combinations (noncrossovers) and two with recombinations of genes (crossovers) which could arise only if two of the four strands exchanged parts between the two loci being studied.

In Neurospora and related organisms all four products of meiosis can be recovered and all four members of a meiotic "tetrad" analysed in detail. The same type of proof that crossing over between two loci involves only two of the four chromatids has been obtained from cases in Drosophila in which some of the products of meiosis can be identified because of chromosome irregularities. In the case of nondisjunction of X chromosomes heterozygous for two mutant genes, it sometimes happens that a crossover chromatid such as *Ab* and a noncrossover chromatid such as *ab* are found in the same nondisjunctional egg, which thus has two X chromosomes. If crossing over had occurred between whole chromosomes, no noncrossover strands would be found; hence their presence indicates crossing over between chromatids. Similar evidence has been obtained from trisomic chromosomes with linked genes in maize.

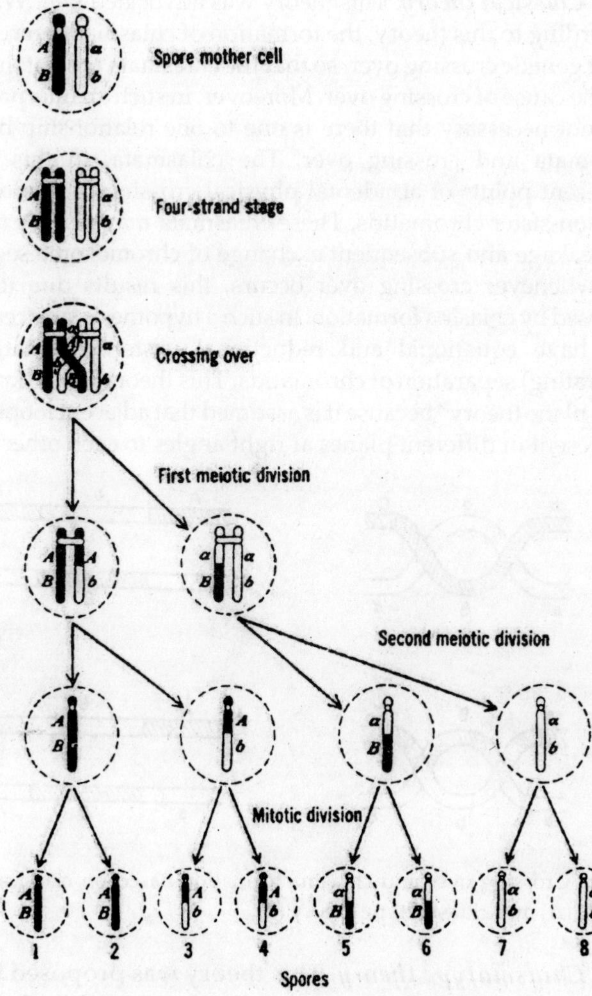

Fig. 4. Scheme of meiosis in Neurospora, illustrated by the behaviour of one pair of chromosomes.

Crossing Over and Chiasmata Formation

There are two theories to explain the relationship between crossing over and chiasmata formation. These two theories will be briefly discussed in this section.

Classical theory. This theory was advocated by L.W. Sharp. According to this theory, the formation of chiasmata precedes the act of genetic crossing over, so that the chiasmata are not the result but the cause of crossing over. Moreover, in such an interpretation, it is not necessary that there is one to one relationship between chiasmata and crossing over. The chiasmata, in this theory, represent points of accidental physical crossing of homologous, but non-sister chromatids. These chiasmata may or may not lead to breakage and subsequent exchange of chromosome segments, but whenever crossing over occurs, this results due to strain imposed by chiasma formation. In such a hypothesis, adjacent loops will have equational and reductional (sister chromatids not separating) separation of chromatids. This theory is also known as "two plane theory" because it is assumed that adjacent loops would be present in different planes at right angles to each other.

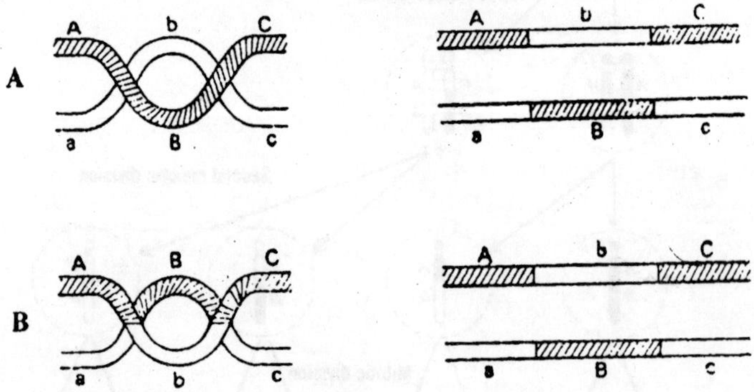

Fig. 5. Crossing over and chiasma formation based on classical theory (A) and chiasmatype theory (B).

Chiasmatype theory. This theory was proposed by F.A. Janssens in 1909. It was later fully developed by J. Belling and C.D. Darlington. According to this theory, the true chiasmata are the direct result of crossing over so that crossing over really precedes chiasmata formation. As opposed to classical theory, in chiasmatype theory, crossing over results from breakage followed by reunion or exchange of non-sister chromatid segments from homologous chromosomes. Since in this theory, chiasmata are the result of crossing over, there would be one to one relationship between the chiasmata and crossing over. This theory is also called 'one plane theory' as against the 'two plane theory' discussed above. In this theory, one would expect reductional separation of chromatids on either side of a chiasma. The classical theory or the two plane theory is now of only historical importance, since all available experimental evidence goes in favour of the 'chiasmatype theory' or 'one plane theory'.

Precocity theory. This theory was discussed by C.D. Darlington. Darlington extended this theory to explain recombination also. The precocity theory presumes that prophase is precocious in meiotic cell division and therefore involves homologous pairing to satisfy the pairing need, which is achieved in mitosis due to replication of the chromosomes. This theory is now untenable in the light of the recent information that DNA synthesis really takes place before the onset of prophase I in meiosis, although synthesis of very small fraction of DNA (0.3 per cent) is delayed till zygotene. Precocity theory presumes that the DNA synthesis or chromosome duplication takes place later in pachytene or diplotene and then results in the separation of homologous chromosomes (terminalisation). On the basis of precocity theory, Darlington explained crossing over to be the result of strain or torsion produced due to coiling of homologous chromosomes and the sister chromatids.

Copy choice hypothesis. The copy choice theory was proposed by J. Lederberg in 1955. In this it is assumed that a newly synthesized daughter chromatid is derived due to copying of one chromosome up to a certain distance and then switching on to the other homologous chromosome for copying the remaining distance or region of the chromosome. Copy choice theory was also inadequate, because it assumed a conservative rather than a semiconservative replication of DNA duplexes. In these copy choice theories (several such theories are known), a chromosome with DNA duplex is synthesized *de novo* by copying the parent chromosome. This is hardly the case, since we know that the chromosomes replicate in a semi-conservative manner. Although recent modifications of copy choice could overcome the above difficult situation, it still remains a complex and unlikely mechanism of recombination.

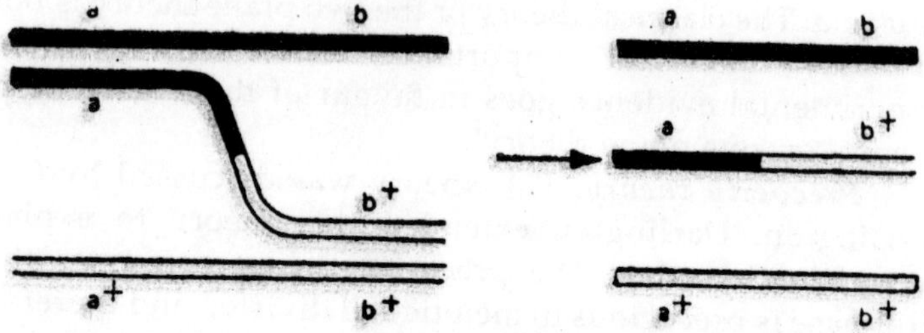

Fig. 6. Mechanism of genetic recombination based on copy choice mechanism.

Hybrid DNA models. Two hybrid DNA models have been proposed, one proposed in 1963 by H.L.K. Whitehouse of Cambridge, England and the other in 1964 by R. Holliday of London. The two models mainly differ in so far as Whitehouse believed that breaks would occur in single strands having opposite polarity, white Holliday proposed that breaks would occur in strands having same polarity. Recently, *'hybrid DNA models'* have become very popular, since these models best explain the results obtained in certain microbial systems. In these models, only one strand in each of two DNA duplexes belonging to nonsister chromatids breaks. The single strands released from these breaks then pair crosswise with unbroken strands by complementary base pairing. This results in the formation of hybrid DNA segments, and hence the name hybrid DNA models. In recent years, the role of a protein Rec A, synthesized under the influence of a gene *rec* has been emphasised. Rec A utilizes single stranded DNA and ATP for its activity and helps single stranded DNA to base pair with its complement to form a hetero-duplex (hybrid DNA), which then leads to recombination at the molecular level.

Fig. 7. Diagram showing the terminalisation.

Break and Exchange Theory

This theory has been originally forwarded by Muller and is a most widely accepted theory about the mechanism of crossing over. It suggests that (i) prior to crossing over, the chromosomes of each bivalent get duplicated to form a tetrad; (ii) crossing over occurs only in non-sister chromatids of a tetrad; and (iii) the crossing over involves the mechanical breaks in non-sister chromatids, mutual exchange of chromosomal segments between the broken non-sister chromatids and reunion or recombination of chromatids during the early part of meiotic prophase I. Most geneticists are agreed with this theory, however, there exists a lot of controversy about the manner and time of the break of chromatids during the crossing over. Different cytologists and geneticists have forwarded different views regarding breakage process of chromatids and these views are following:

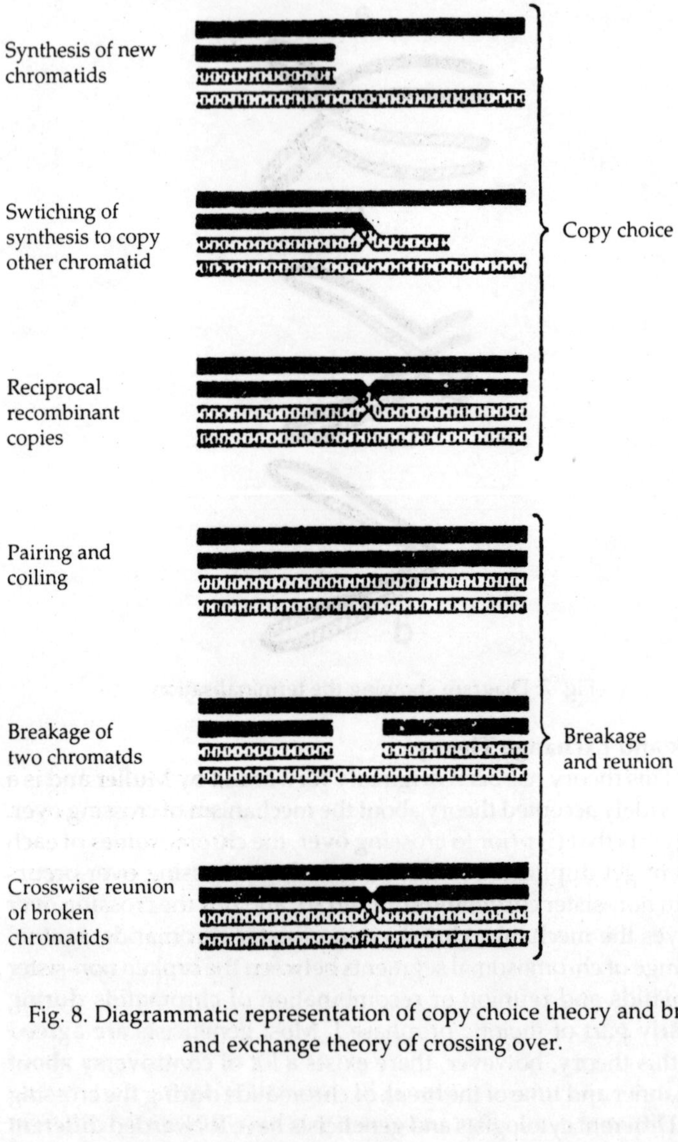

Fig. 8. Diagrammatic representation of copy choice theory and break and exchange theory of crossing over.

Contact first theory. This theory was proposed by Serebrovsky. According to this theory two chromatids of a tetrad first touch and cross each other and then at the point of cross the breakage of chromatids occur. The chromsome segments then unite to form new combinations.

Strain or torsion theory. Darlington (1935) propounded that chromosomal breaks during crossing over occur as a result of strain during pairing. According to this theory during meiotic prophase I when homologous chromosomes twist around each other in synapsis, then each of the two pairing chromosomes, itself becomes spirally woundup. This is called *spiralisation.* The chromosomes are thus under heavy strain due to tight coiling or twisting, and are capable of bearing this strain so long as they are not split. After splitting, each chromosome forms two weaker chromatids, so that one chromatid breaks at one point. This releases the strain on the sister chromatid but somehow, it increases the strain on the non-sister chromatids causing a break in one of them at a point opposite to the first break. The broken chromatids are now unwind near their breaking points and during this, they come into contact and unite, thus forming a chiasma.

Differential contraction theory. Huskins and Newcombe (1941) suggested that tension set up by differential contraction after pairing may cause breaks at overlaps. This hypothesis is based on measurements which indicate that contraction is greater in paired regions than in those not yet paired.

White's frontier theory. White (1951) suggested that the chromosome does not split simultaneously along its entire length. Further, he suggested that the unsplit parts of both homologous chromosomes while remain together by pairing force, the split regions of chromosomes are being forced apart, by some kind of repulsion and thus, consequently a localised strain is developed at the meeting point of the split and unsplit regions, causing a breakage in the chromosome.

Breakage first theory. This theory was proposed by Muller. It holds that before the occurrence of chiasma and crossing over two chromatids break in two fragments and then reunion of chromosomal segments occur in new arrangement.

Absence of Crossing over in Drosophila Males

In most of the organisms that have furnished the classical materials for genetic studies, such as maize, peas, mice, poultry, man and others, recombination of linked genes takes place both in females and in males. Similarly, in most organisms favourable for cytological studies, formation of chiasmata is observed in both female and male meiosis. However, the most thorough and extensive studies on linkage and crossing over have been made on species of Drosophila, and it happens that in these insects the situation is very different in the two sexes. In Drosophila, crossing over rarely or never takes place in the male.

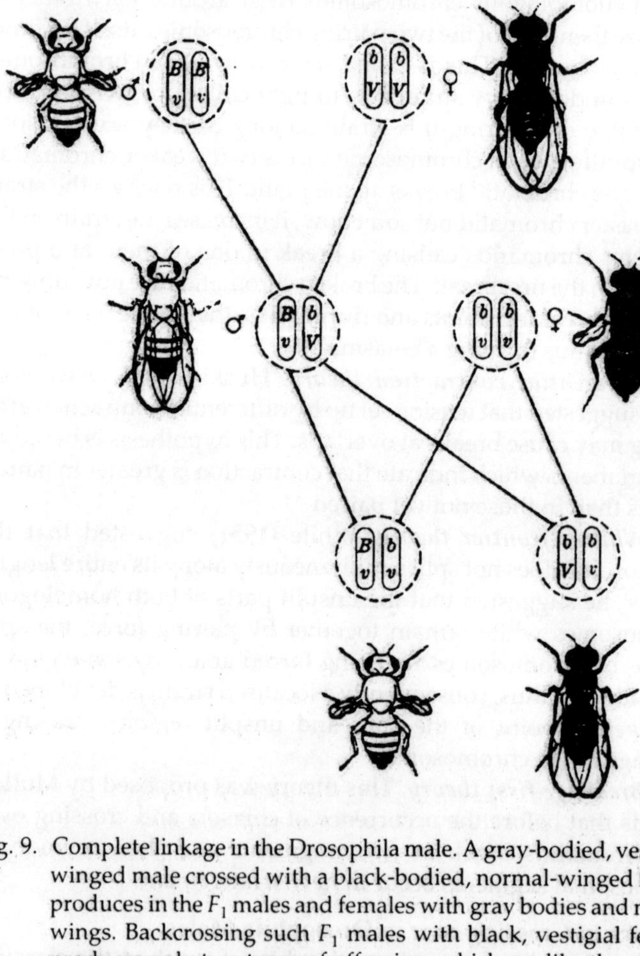

Fig. 9. Complete linkage in the Drosophila male. A gray-bodied, vestigial-winged male crossed with a black-bodied, normal-winged female produces in the F_1 males and females with gray bodies and normal wings. Backcrossing such F_1 males with black, vestigial females produces only two types of offspring, which are like the original parents. There are no recombinations.

If a gray-bodied vestigial-winged fly is crossed to a black-bodied long-winged one, the F_1 generation consists of gray long-winged (normal) flies. Thus the gene for gray body colour, *B*, is dominant over its allele, which causes black body colour, *b:* and the gene for long-wings, *V*, is dominant over its vestigial allele, *v*. Now, if the F_1 male hybrids are crossed to double recessive females (black-bodied vestigial-winged females), only two kinds of offspring are produced: gray vestigial and black long. The expected types of crossovers—gray long and black vestigial—do not appear at all. If, however, an F_1 female fly is crossed with a black vestigial male, the four expected types are produced in the following proportions:

Noncrossovers

Gray vestigial	Black long
41.5 per cent	41.5 per cent

83 per cent

Crossovers

Black vestigial	Gray long
8.5 per cent	8.5 per cent

17 per cent

Recombination is evident in about 17 per cent of the gametes. The experiment shows that a perceptible distance separates the genes for black and vestigial and that absence of crossovers in the gametes of the male is not due to the extreme closeness of the genes in the chromosome. It must, therefore, be due to conditions peculiar to the male, such as the nonoccurrence of chiasmata in spermatogenesis.

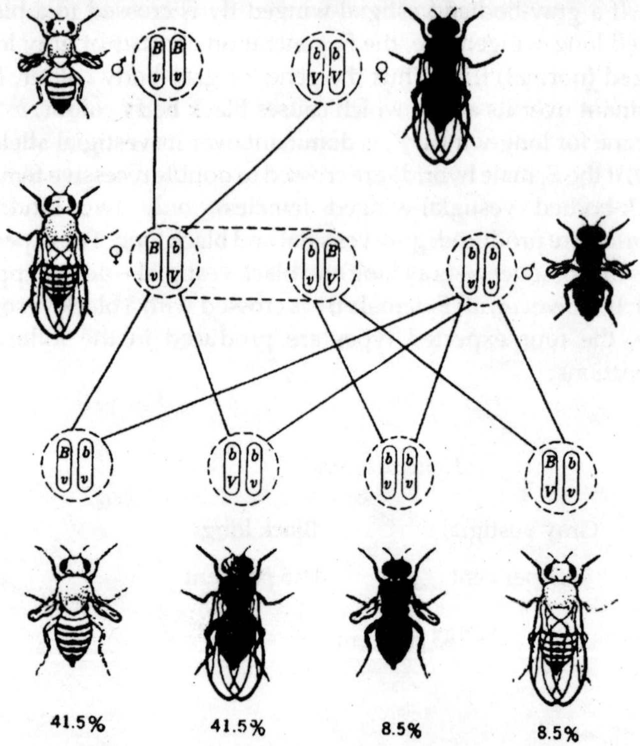

Fig. 10. Recombination of linked genes in the Drosophila female. Results of a cross similar to that in Figure 9 but with the F_1 females backcrossed with black, vestigial males.

Only a few cases are known in which linkage is complete in one sex and crossing over occurs more or less freely in the other. In all species of Drosophila so far studied in this respect, crossing over is absent or exceedingly rare in males, whereas in the silk-worm moth it seems to be absent in females. Cytological study of spermatogenesis in male Drosophila, made by Darlington and others, discloses that the homologous chromosomes undergo, as usual, pairing in the spermatocytes; however, no chiasmata are established, at least not in the autosomal bivalents. The bivalents eventually come to consist of four chromatids lying parallel to one another. At the first meiotic division, pairs of chromatids go to each pole, and at the second meiotic division, single chromatids pass to each cell and the cells are transformed into spermatozoa. Female meiosis in Drosophila is unfavourable for detailed cytological study, but it appears certain that the bivalents do have chiasmata.

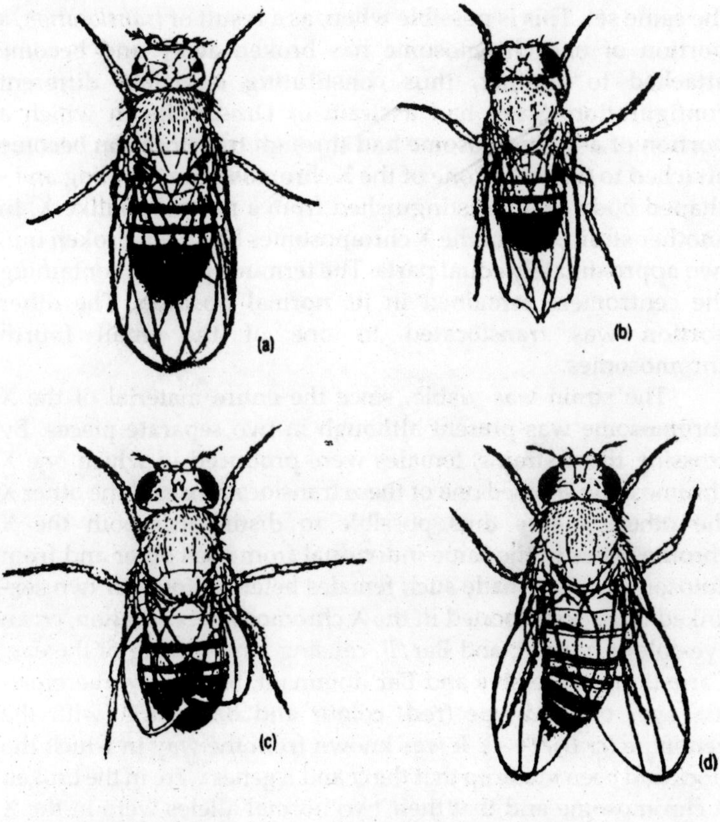

Fig. 11. Mutants of *Drosophila melanogaster*: (a) Bar eyes; (b) cut wings; (c) rudimentary wings; (d) rotated abdomen.

Stern's Experiment

Further evidence of association of genes with chromosomes is provided by the demonstration that, where there is an interchange of material between two homologous chromosomes, there is an interchange of genes by crossing over. This has been shown in a number of cases, but Stern's demonstration in Drosophila is simple and complete. The essential feature of this demonstration is the use of strains in which it is possible to distinguish under a microscope the two members of a pair of chromosomes from each other and from other chromosomes of the same set. This is possible when, as a result of *translocation,* a portion of one chromosome has broken away and become attached to another, thus constituting a visibly different configuration. Stern had a strain of Drosophila in which a portion of a Y chromosome had through translocation become attached to the end of one of the X chromosomes, forming an L-shaped body easily distinguished from a normal rodlike X. In another strain, one of the X chromosomes had been broken into two approximately equal parts. The terminal portion containing the centromere remained in its normal position. The other portion was translocated to one of the small fourth chromosomes.

The strain was viable, since the entire material of the X chromosome was present although in two separate pieces. By crossing these strains, females were produced in which one X chromosome showed one of these translocations and the other X the other; it was thus possible to distinguish both the X chromosomes in the same individual from each other and from autosomes. Stern made such females heterozygous for two sex-linked mutations located in the X chromosome: carnation, *cr,* an eye-colour mutant, and Bar, *B,* causing a narrowing of the eye. Carnation is recessive and Bar dominant. Such a fly, therefore, has eyes of wild-type (red) colour and Bar shape, with the genotype (*cr B*)/(+ +). It was known from the way in which the stock had been made up that the *cr* and *B* genes were in the broken X chromosome and that their two normal alleles were in the X bearing the translocated Y portion. Such a female was bred to a male having both the recessive genes (*cr* and the wild-type allele of Bar) in its X chromosome. The offspring of such a cross of double heterozygote by double recessive could be classified by inspection into four groups, as in any case of linkage. The females alone were studied.

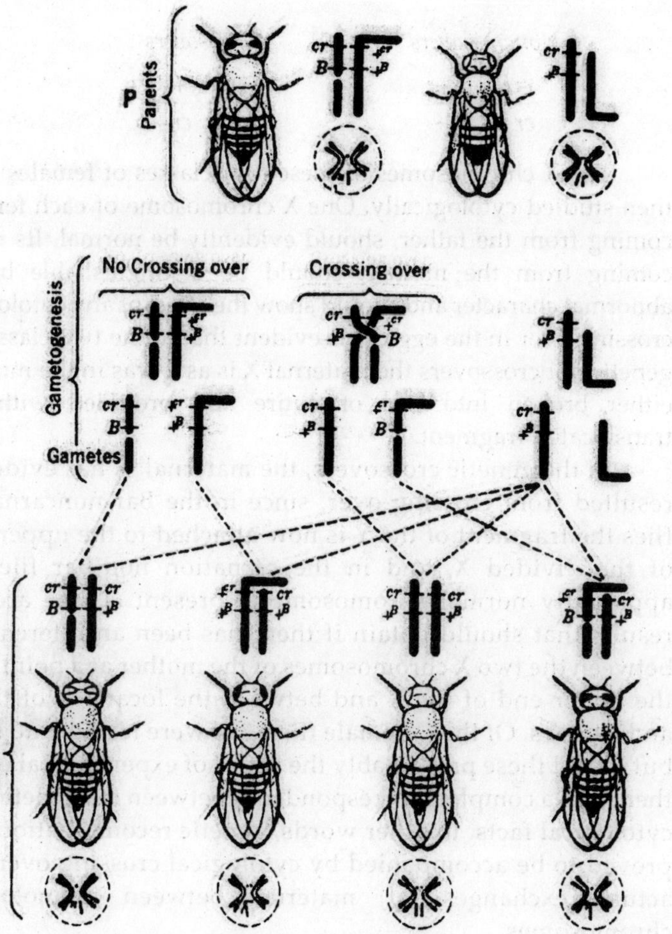

Fig. 12. Diagram of recombination of the genes for Bar (B, narrow eyes) and carnation (*cr*, an eye colour) in females in which the two X chromosomes are distinguishable in appearance under the microscope.

Of these there were two noncrossover classes (carnation Bar and normal) and two crossover classes (carnation with normal shape and Bar with normal colour). These may be represented as follows:

Noncrossovers		*Crossovers*	
$\frac{cr B}{cr +}$	$\frac{+ +}{cr +}$	$\frac{cr +}{cr +}$	$\frac{+ B}{cr +}$

The X chromosomes of these four classes of females were then studied cytologically. One X chromosome of each female, coming from the father, should evidently be normal. Its mate, coming from the mother, should be distinguishable by its abnormal character and should show the effect of any cytological crossing over in the eggs. It is evident that in the two classes of genetic noncrossovers the maternal X is as it was in the mother, either broken into two or entire and provided with the translocated fragment.

In the genetic crossovers, the maternal X has evidently resulted from crossing over, since in the Bar noncarnation flies the fragment of the Y is now attached to the upper half of the divided X, and in the carnation non-Bar flies an apparently normal chromosome is present. These are the results that should obtain if there has been an interchange between the two X chromosomes of the mother at a point near the upper end of the X and between the locations of the *cr* and *B* genes. Of the F_1 female flies, 364 were tested, and in all but 5 (and these presumably the result of experimental error) there was a complete correspondence between the genetic and cytological facts. In other words, genetic recombination was proved to be accompanied by cytological crossing over and actual exchanges of material between homologous chromosomes.

McClintock's experiment. Creighton and McClintock like Stern made convincing correlation between cytological evidence and genetical results of crossing over in maize. They made the use of knob of 9th chromosome of maize which in two different strains might had one allele for either coloured aleurone (C) or colourless aleurone (c) and one allele for either starchy endosperm (Wx) or waxy endosperm (wx).

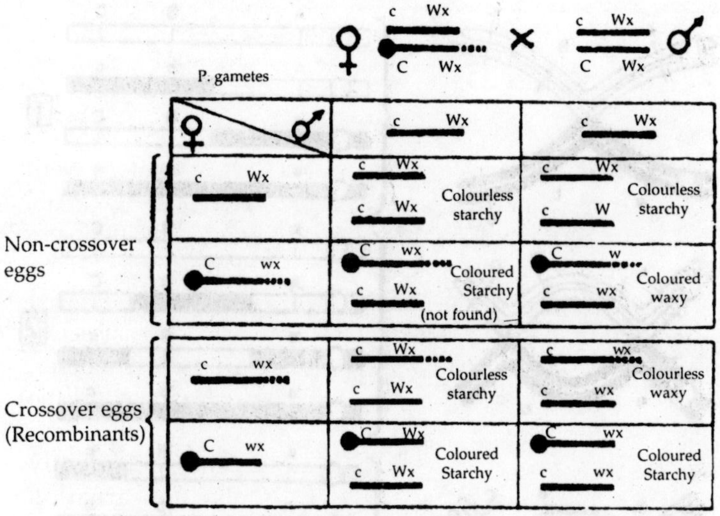

Fig. 13. A diagrammatic representation of parallelism between cytological and genetic crossing over showing the phenotypes of aleurone and endosperm of F_1 grains and chromosome morphology and genotypes for microsporocytes produced by F_1 plants.

Kinds of Crossing Over

According to the number of chiasma the crossing over may be of following kinds:

Single crossing over. When only one chiasma occurs only at one point of the chromosome pair, it is called **single crossing over.** It produces two non-crossover chromatids and two cross over chromatids.

Double crossing over. When the crossing over occurs at two points between any two given points in the same chromosome pair, it is called *double crossing over.* It produces four crossovers. In double crossing over following two types of chiasma may be formed.

(A) Reciprocal Chiasma. In reciprocal chiasma same two chromatids are involved in the second chiasma as in the first. It produces only two non-crossover chromatids, because, it restores the order which was changed by the first chiasma. In this chiasma

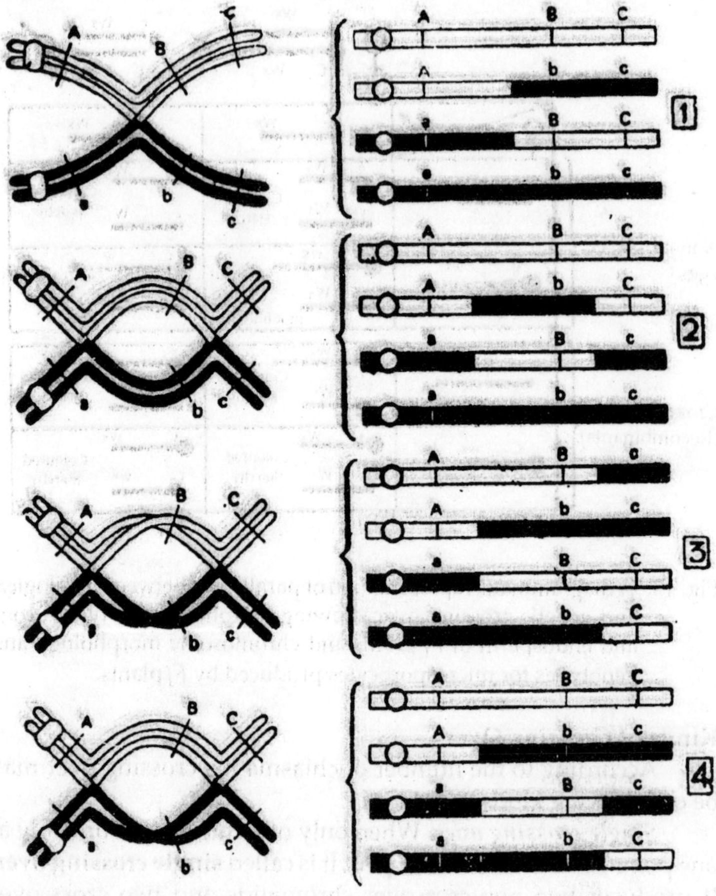

Fig. 14. Diagram showing the results of single and double chiasma. 1. Single chiasma. 2. Two strand double cross over. 3. Four strand double cross over. 4. Three strand double cross over.

(B) Complimentary chiasma. When both the chromatids taking part in the second chiasma are different from those chromoatids involved in the first chiasma, the chiasma is called *complimentary chiasma.* It produces four single crossovers but no non-cross over. In complimentary chiasma all the four chromatids of a tetrad are involved.

Multiple crossing over. When crossing over occurs at three, four, or more points between any two given points in the same chromosome pair these are called *triple, quadruple* or *multiple crossing over*.

Factors Affecting Crossing Over

Many extrinsic and intrinsic factors have been found to affect the rate of crossing over. Among these are the effects of sex, age, temperature, water content, ionizing radiation, certain chemicals, proximity to the centromere or heterochromatic regions, chromosomal aberrations such as inversions, etc. For example, in certain species, crossing over may be abolished or depressed in one sex, as have been tabulated in the following table:

Table 1
Sexual Differences in Crossing Over in Different Species

S. No.	*Name of Organism*	*Male*	*Female*
1.	Garden pea	present	equal to male
2.	*Drosophila*	absent	present
3.	Silkworm	present	absent
4.	Mouse	present	higher than male
5.	Pigeon	higher than female	present

Gowen 1933 have shown in that *Drosophila* mutation suppressed crossing over completely. Bridges (1927) demonstrated that as *Drosophila* aged, crossing over was depressed particularly between genes which lay close to the centromere. Likewise, Plough (1917) showed that at temperature between 17°C and 29°C the rate of crossing over remains constant, but at below or above this temperature range, the rate of crossing over increases. According to Sybenga (1972) low water contents can lead to reduced chiasma frequencies; in extreme cases it can result in complete failure of pairing and exchange. Further, it is well known fact that the rate of crossing over is reduced in heterochromatic regions of chromosomes.

Ramanna (1971) has observed frequent crossing over in the entirely heterochromatic short arm of chromosome 2 of pollen mother cells of tomato. Ionizing radiation increase the rate of mitotic as well as meiotic crossing over in female *Drosophila*. In male *Drosophila* too, crossing over can be induced by ionizing radiation. Lawrence (1965) has reported that irradiation of *lilium, Tradescantia* and *Chlamydomonas* at specific moments before and during prophase of meiosis strongly affected chiasma frequency and crossing over. Irradiation at premeiotic DNA synthesis had a reducing effect and irradiation at prophase (zygotene-pachytene) had an enhancing effect. Certain chemicals such as ethyl-methane-sulphonate and radiomimetic substances have been found to increase somatic crossing over in years. Colchicine prevents chromosome pairing and in this way reduced crossing over. Selenium, in its excess amount has a reducing effect on crossing over.

Significance of Crossing Over

The process of crossing over has following significance:

1. The frequency of crossing over is of great use in constructing genetic maps of the chromosomes.
2. It provides direct evidence for linear arrangement of linked genes in chromosomes.
3. It increases the frequency of genetical variations which are the raw materials of organic evolution.

Chromosomal Mapping

Morgan and his fellow workers made numerous crosses involving linked genes. It was found that, in successive experiments, the amount of crossing over between two particular genes was always the same. Depending on the genes used, the amount of crossing over might be less than 1 per cent or nearly 50 per cent. Morgan suggested that the different values were the result of the relative positions of the genes on the chromosome: If the amount of crossing over between hypothetical genes *A* and *B* was small and beween genes *C* and *D* large, one would predict that *A* would be closer to *B* than C would be to *D*. The development of this concept, that linkage data could be used to map the relative positions of the genes on the chromosomes, was attempted in 1913 by Sturtevant, a student of Morgan.

Sturtevant made crosses involving five genes carried on the X chromosome, namely, yellow body (*y*), *white* eyes (*w*), *vermilion* eyes (*v*), *miniature* wings (*m*), and *rudimentary* wings (*r*). From the data obtained, he constructed a genetic map showing the 'positions' of these genes on the **X** chromsome. This was his basic assumption: 'It would seem . . . that the proportion of "crossovers" could be used as an index of distance between any two factors. Then by determining the distances (in the above sense) between A and B and between B and C, one should be able to predict [the amount of crossing over in the interval] AC. For, if proportion of crossovers really represents distance, AC must be approximately, either AB plus BC, or AB minus BC, and not any intermediate value.'

The percentage of crossovers between *y* and *v* was found to be 32.2 and between *y* and *m* 35.5. On the basis of the hypothesis we would expect *v* to be closer to *y* than *m* to *y*. What can we conclude about the relative positions of *m* and *v*. According to Sturtevant this should be 67.7 (35.5 + 32.2) or 3.2 (35.5 – 32.2). The reason for this is as follows:

The chromosome is known to be a very long and very thin structure. We can consider it as having length only, and can represent it as a long line. On this line we shall put *y* and *v* as follows:

y *v*

-----3.2------

Now *m* can be either to the right or to the left of *y*, as shown here:

? *m* *y* *v* ? *m*

-----35.5------ -----32.2------

-----35.5------

If *v* and *m* are on the same side of *y*, we would expect the amount of crossing over between *v* and *m* to be 3.2 per cent. If they are on opposite sides of *y* the value should be 67.7 per cent. When Sturtevant measured the amount it was found to be 3 per cent, which indicated that *v* and *m* were on the same side of *y*. This close correspondance between the actual and expected result was strong support for his theory.

In this manner, the relative positions of the five genes were determined and a genetic map constructed. The *y* gene was taken as the reference point and the distances measured from it. This was the result.

y	*w*	*v*	*m*	*r*
0.0	1.0	30.7	33.7	57.6

It was found that the most reliable values were obtained when crossover values for adjacent genes were used. This was due to the occurrence of double crossover, which introduced an error into the results. Let us consider the three genes, *y, w* and *m*. Sturtevant raised 10,495 flies to test the linkage relations. He found that in 6,972 flies there was no crossing over between the three genes. In 3,454, crossovers occurred between *w* and *m;* in 60, crossovers between *y* and *w* were detected: and in nine a double crossover occurred. That is, there was one crossover between *y* and *w* and another between *w* and *m*. This would result in *y* and *m* being on the same chromosome as they were before the two crossovers occurred.

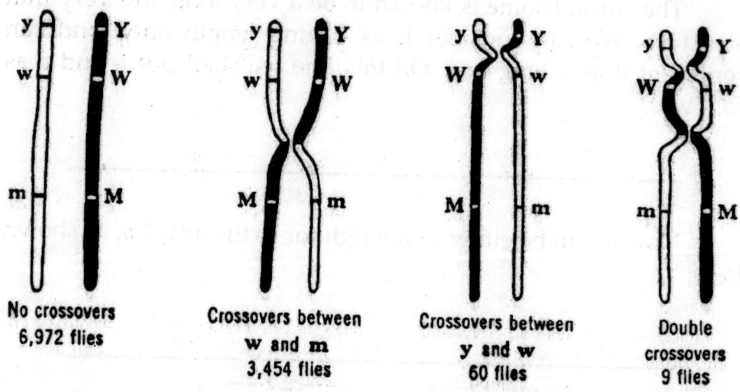

Fig. 15. B 11-1 Sturtevant's Experiment.

In the case of the double crossovers it should be noted that three genes are always necessary to detect the event. Thus, if only the genes *y* and *m* were used, any double crossover between them would be undetected since *y* and *m* would still be together after the chromatids had broken and rejoined. When genes are far apart, double crossovers are likely. If they are not detected they will introduce an error in the positions assigned to the genes, for the data would indicate that the genes are closer to one another than they really are. For this reason Sturtevant suggested that chromosome maps be based on crossover values of genes close to one another and not those far apart.

Now the question arises, What is the relation of the chromosome map to the position of these genes on the chromosome? Sturtevant has this to say:

> Of course, there is no knowing whether or not these distances as drawn represent the actual relative spatial distances apart of the factors. Thus the distance *wv* may in reality be shorter than the distance *yw*, but what we do know is that a break is far more likely to come between *w* and *v* than between *y* and *w*. Hence, either *wv* is a long space, or else it is for some reason a weak one. The point I wish to make here is that we have no means of knowing that the chromosomes are of uniform strength, and if there are strong or weak places, then that will prevent our diagram from representing actual relative distances—but, I think, will not detract from its value as a diagram.

Sturtevant's chromosome map was a graphic way of expressing linkage data. Once constructed, these maps proved useful in predicting the results of untried crosses. The most important induction from the data is that the genes are arranged in a linear order on the chromosomes, analogous to the sequence of beads on a string. The position occupied by a gene is its *locus*.

6

Sex Determination

In higher animals, sexes are usually separate with or without sexual dimorphism. Their identification depends upon the secondary sexual characters. However, the determination of sex deals with the knowing of sex before its differentiation.

The first impression cytologists gained from a study of chromosomes was a feeling that the behaviour of these cell structures was similar in all animals. Whether the form studied was a worm, snail, salamander, or mammal, one observed the same duplication of each chromosome to form two chromatids during mitosis. This was followed by the distribution of one of the two chromatids to each daughter cell. Each chromosome appeared double at the same time, and the movements during metaphase, anaphase and telophase were synchronous. A similarity of behaviour was observed in the events connected with the reduction of chromosome number in meiosis of both male and female gametes. It should be emphasized that this similarity of behaviour applied not only to the behaviour of the group of chromosomes but to the individual chromosomes in the group as well.

Atypical Chromosomes. It was not long before this concept of uniform behaviour was found to have its exceptions. In the last decade of the nineteenth century and the first years of the twentieth, a few cases were reported or one or two of the chromosomes of a set behaving in a manner quite unlike the rest. The unusual behaviour might be a difference in reaction of the chromosome to stains, meiotic movements that were not synchronous, or the presence of 'extra' or 'accessory chromosomes.'

The term *accessory chromosome* referred to those cases where there was one chromosome without a mate, in contrast to the usual situation of all chromosomes being in morphologically similar pairs. The analysis of accessory chromosomes led to important results concerning the role of chromosomes in heredity. As is frequently the case in science, these observations were made and recorded before their true significance was understood.

The Sexual Function of X and Y Chromosomes

These facts concerning sex determination were discovered early in the current century by Wilson, Sutton, Stevens, Montgomery, and others. Many questions were, however, left unsettled. One of the most obvious is: What are the functions of the X and Y chromosomes in determining the sex of Drosophila? Is the presence of a Y chromosome essential in the male? Morgan's investigations of sex-linked inheritance showed that many sex-linked genes are carried in the X, whereas the Y behaved as though it were "empty" of genes. The critical evidence was supplied by Bridges in 1916. Drosophila males were discovered which lack a Y chromosome (XO males). Such males are sterile but otherwise normal in appearance and in behaviour. Females may carry a Y chromosome (XXY females) and, even so, be normal and fertile. This seemed to show that, at least in Drosophila, the presence of one X chromosome makes a male and of two X's a female. The Y chromosome is sexually "neutral."

Table 1
Varieties of Chromosomal Sex Determination

Organism	*Heterozygous Sex*	*Gametes*		*Zygotes*	
		Sperm	*Eggs*	*Females*	*Males*
Drosophila, man, etc.	Male	X and Y	All X	2X = ♀	XY = ♂
Protenor (bug), grasshoppers	Male	X and O	All X	2X = ♀	X = ♂
Birds, moths	Female	All X	X and Y	XY = ♀	2X = ♂
Fumea (a moth)	Female	All X	X and O	X = ♀	2X = ♂

Sex Chromosome Mechanisms

In sexually dimorphic dioecious organisms besides morphological and behavioural differences between both sexes, the sexual diversity also occurs at the level of chromosomes. The chromosomal differences between the sexes of several dioecious species were found earlier in the course of cytological investigations. A German biologist, Henking in 1891 while studying spermatogenesis of the squash bug, *pyrhocoris,* noted that half of the spernatozoa contained an extra chromosome which was called by him as "X body". The significance of X body was not immediately understood, but in 1902 an American geneticists McClung who had made extensive observations of gametogenesis in grass hoppers, suggested that the X body was involved in some way with the determination of sex. He reported that the somatic cells of the female grasshopper *(Xiphidium fasciatum)* contained 24 chromosomes whereas those of the male had only 23. Three years later, Miss Steven and Wilson succeded in following both oogenesis and spermatogenesis in certain bugs of *Protenor* genus. In these insects, different number of chromosomes were observed in the cells of two sexes, the diploid number in the female being 14 while that of the male was 13. By following oogenesis, these investigators found that all the eggs contained 7 chromosomes. A similar study of spermatogenesis showed that two types of sperms were formed, those with 7 chromosomes and those with 6 chromosomes. During these studies it was also realised that the X body was a chromosome, so the X body became known as X chromosome. Because the X chromosome has been found to have relation with determination of sex of individual by its presence or absence so it is called *sex chromosome.*

Stevens and Wilson (1905) while working with the milkweed bug, *Lygaeus turcicus* found same number of chromosome in both the sexes, (i.e., 14). In female all chromosomes were paired and the homologous were equal in size. In the male, all the chromosomes were paired, but the chromosome identified as the homologue to the X chromosome was distinctly smaller and was called the *Y chromosome.* In dioecious organisms thus two types of chromosome were recognised which are as follows:

Autosomes: The chromosomes which have no relation with the sex and contain the genes which determine the somatic characters of the individuals are known as *autosomes* (A).

Sex chromosomes: The chromosomes which are responsible for the determination of sex are known as *sex chromosome*, e.g., X and Y chromosomes.

Theories of Sex Determination

Several theories have been proposed to explain the mechanism of sex determination. Some important theories are being described here:

Theory of Heterogamy. This theory was proposed by *Henking* and *Mc Clung.* This theory is applicable in those cases where any of the two parents produces atleast two types of gametes. In some cases males are heterogametic while in few others females are heterogametic as given under the following cases:

(a) Heterogametic Males. In such cases females produce only one type of ova, each bears one X chromosome but males produce two types of sperms, 50 per cent with chromosome and remaining 50 per cent with Y as without any sex chromosomes, as shown in the following cases:

(i) *XX-XO type.* In certain insects specially those of the orders Hemiptera (true bugs) and Orthoptera (grasshoppers and roaches), the female having two X chromosomes (hence referred as XX) and are thus homogametic, while the male having only one X chromosome (hence referred as XO). The presence of an unpaired X chromosome determines the masculine sex. The male lacking in one X chromosome produces two types of sperms: half with X chromosome and half without X chromosome. The sex of the offspring depends upon the sperm that fertilizes the egg (each of which carries a single X chromosome).

(ii) *XX-XY type.* In man, other mammals certain insects including *Drosophila* the female possesses two homomorphic X chromosomes in their body cells (hence referred as XX) and they being homogametic, produce, one kind of eggs, each with one X chromosome. The males of these organisms possess one X chromosome and one Y chromosome (hence referred as XY). The males having two heteromorphic sex chromosomes produce two kinds of sperm: half with X chromosome and half with Y chromosome. The sex of embryo depends on the kind of sperm. An egg fertilised by a X-bearing sperm, produces a female, but, if fertilised by a Y bearing sperm, a male is produced.

1. Sex is determined at the time of fertilisation.
2. If sex determination is due only to sex chromosomes, we can regard the sex of an individual as irreversible, unless we can alter the chromosomes.

If these observations on sex chromosomes are correct, we may draw some important conclusions:

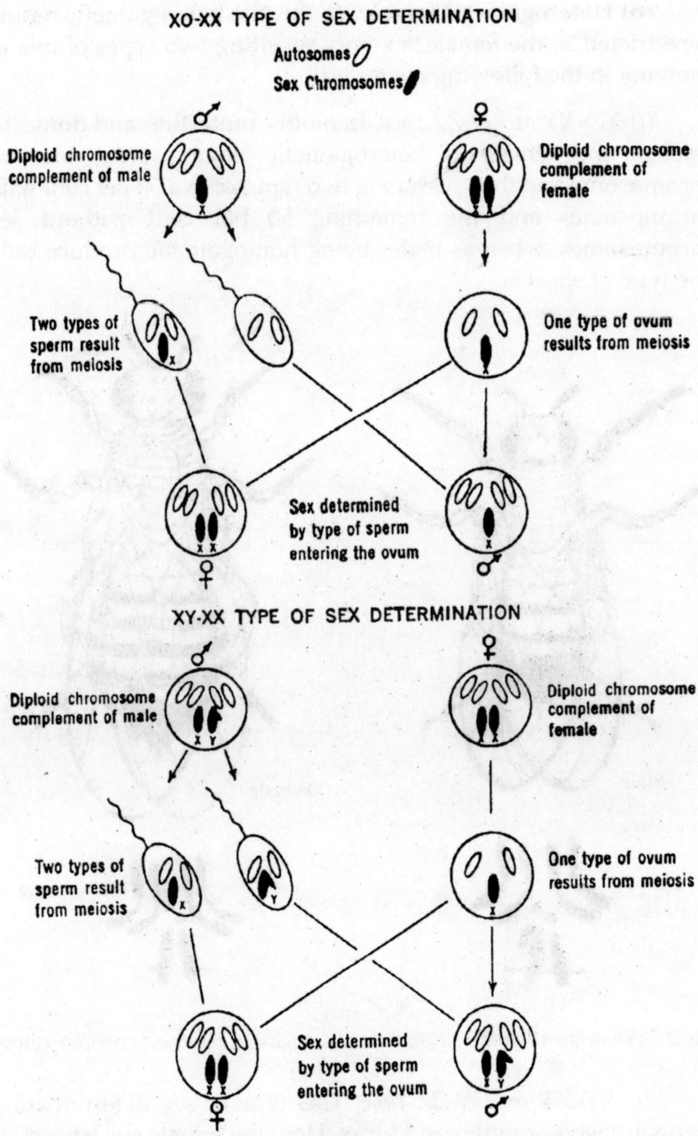
XO-XX TYPE OF SEX DETERMINATION
Autosomes
Sex Chromosomes
Diploid chromosome complement of male
Diploid chromosome complement of female
Two types of sperm result from meiosis
One type of ovum results from meiosis
Sex determined by type of sperm entering the ovum
XY-XX TYPE OF SEX DETERMINATION
Diploid chromosome complement of male
Diploid chromosome complement of female
Two types of sperm result from meiosis
One type of ovum results from meiosis
Sex determined by type of sperm entering the ovum

Fig. 1.

(b) Heterogametic female. In this the heterogametic nature is restricted to the female sex only resulting two types of ova as showing in the following examples.

(i) XO-XY or ZO-ZZ type. In moths, butterflies and domestic chicken the female is heterogametic having only one sex chromosome and thus poducing two types of ova 50 per cent with chromosomes and the remaining 50 per cent without sex chromosomes, whereas males being homogametic produce only one type of sperms.

A

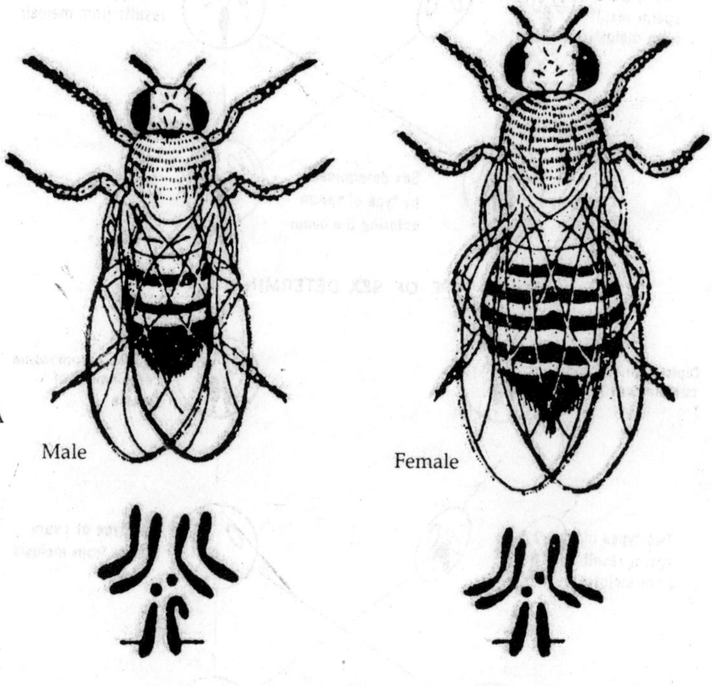

Fig. 2. A male and female *Drosophila melanogaster;* B—their chromosomes.

(ii) XY-XX or ZW-ZZ type. This type of sex determination occurs in fishes, reptiles and birds. Here the female sex has one X or Z and one Y and W chromosome. This is the heterogametic stage resulting two types of ova with X or Z and with Y or W

chromosome. Males are homogametic and produce only one type of sperms.

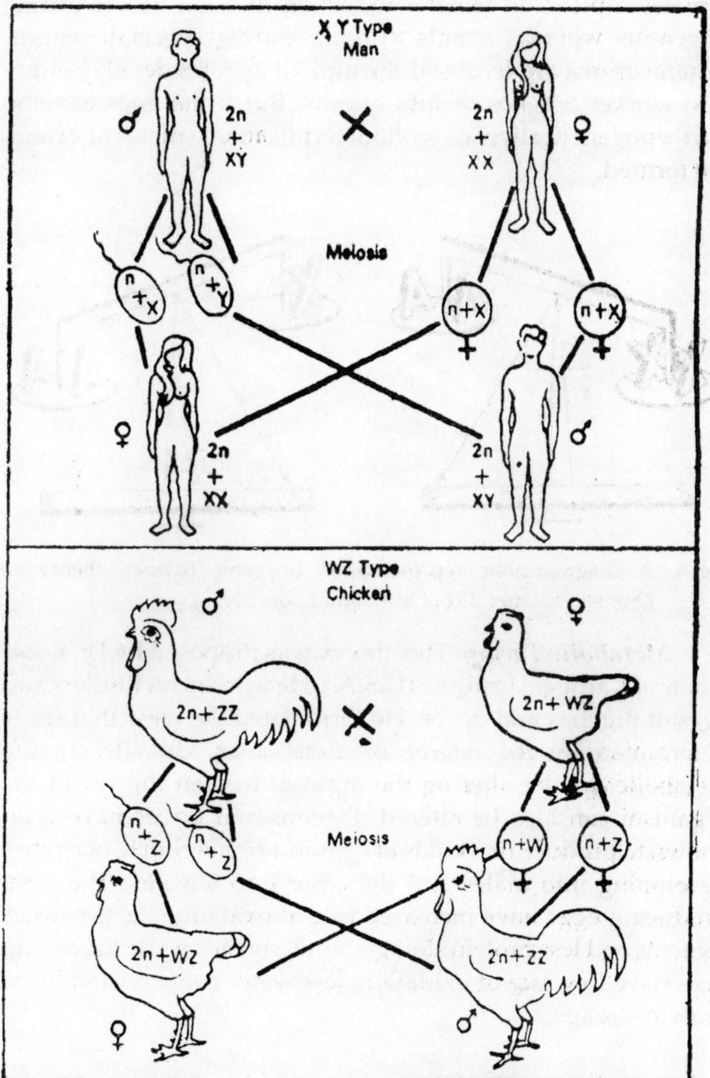

Fig. 3. The XX-XY and ZW-ZZ type determination of sex in man and chicken, (after Cockrum and McCauley, 1965).

(c) Haploid-Diploid type. In honey bee, wasp, ants and certain other hymenopterans males are usually haploid and females diploid. In females reduction division occurs during oogenesis whereas it fails to occur during spermatogenesis. Wherever ova are fertilised the diploid zygotes develop either into worker females or into queens. But if the eggs develop parthenogenetically, i.e., without fertilisation—males or drones are formed.

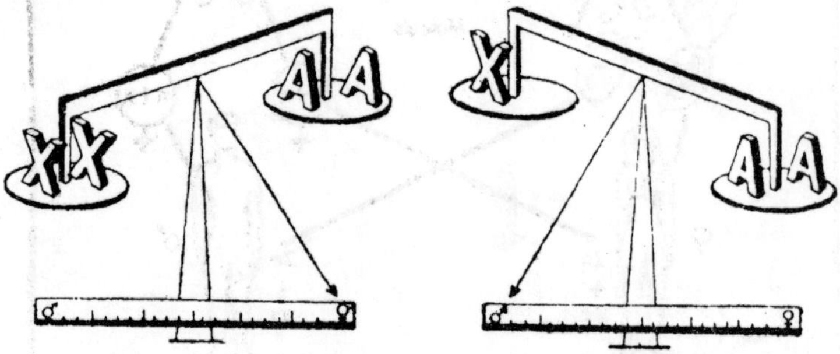

Fig. 4. A diagrammatic representation of genic balance theory in *Drosophila.* (after SRB, Owen and Edgar, 1965).

Metabolic Theory. This theory was proposed by Dr. *Oscar Riddle* of Carnegie Institute (U.S.A.). He worked on Rotifers and eggs of pigeons and doves. He formulated the view that sex is determined by the degree of metabolism. By altering the metabolic rate by altering the input of oxygen the sex of the organism can also be altered. Pigeons and doves have been known to produce two kinds of eggs in a single clutch, one types developing into males and the other into females. The male producing eggs have increased rate of oxidation, larger water contents and less protein storage. Similarly the female producing eggs have low rate of oxidation, less water contents and more protein storage.

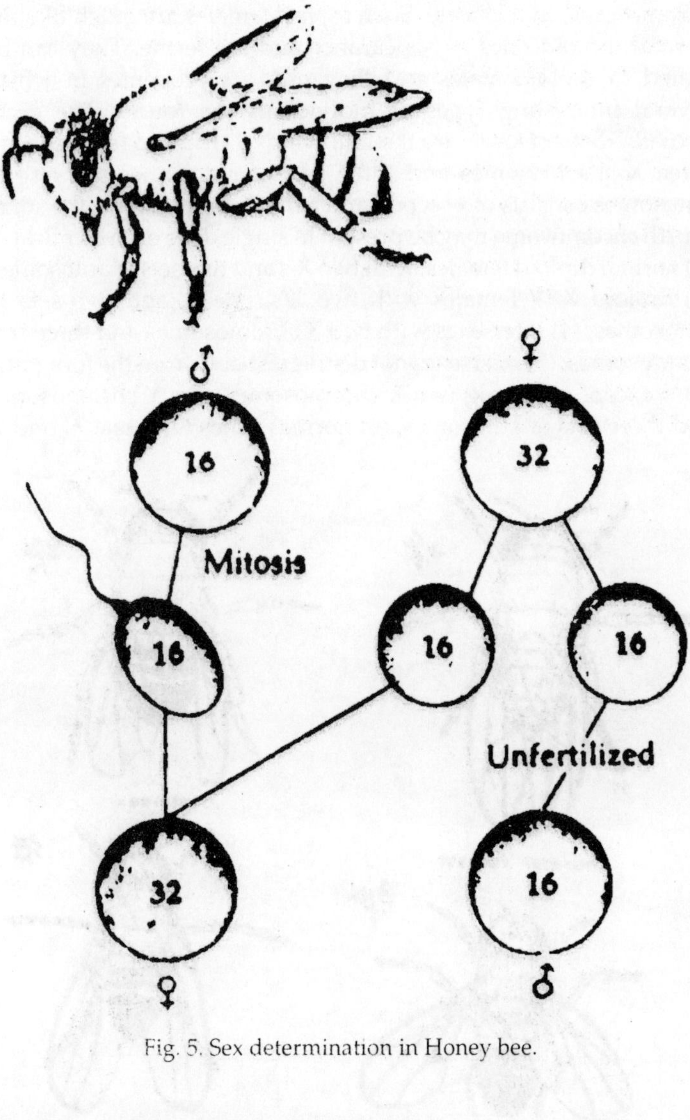

Fig. 5. Sex determination in Honey bee.

Intersexes and Supersexes in Drosophila

An important development of the theory of sex determination took place in 1922. Bridges found some females of *Drosophila melanogaster* that had every chromosome, including the X chromosome, in triplicate. Such *triploid* females are much like the normal diploid ones in appearance and are fertile. They can be mated to diploid males and they produce progenies in which several interesting types of individuals are found. The eight sexually distinct kinds are the following: (1) triploid females with three X chromosomes and three sets of autosomes (a "set" of autosomes consists of a second and a third chromosome; the small fourth chromosome may be present in single dose or in duplicate); (2) normal diploid females, with two X's and two sets of autosomes; (3) diploid XXY females with two X's, one Y, and two sets of autosomes; (4) intersexes with two X chromosomes and three sets of autosomes; (5) intersexes not distinguishable from the foregoing class except in having two X chromosomes, one Y chromosome, and three sets of autosomes; (6) normal males with one X, one Y,

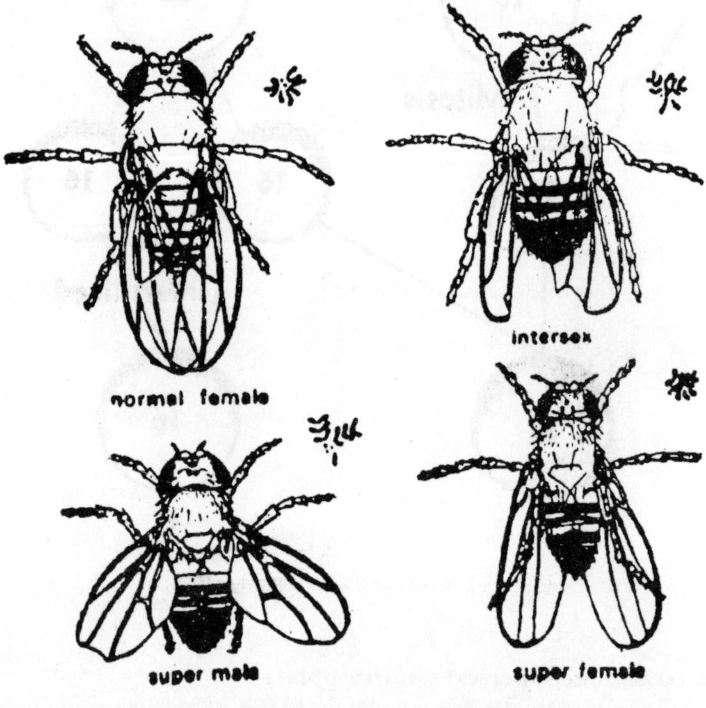

Fig. 6. Sex types in Drosophila.

and two sets of autosomes; (7) superfemales with three X's and two sets of autosomes; (8) superfemales with one X, one Y and three sets of autosomes.

Intersexes are sterile individuals intermediate between females and males. They should not be confused with gynandromorphs, which are individuals having some parts of the body pure female and others male. Superfemales and supermales are also sterile individuals that show some relatively minor differences from ordinary females and males.

Bridges gave the following interpretation to his findings: "The intersexes and 3n (triploid) types lead to the conclusion that sex in *Drosophila melanogaster* is determined by the autosomes as well as by the X chromosomes, the ratio of autosomes to X's being the significant relation. The old formulation of 2X = O is at once seen to be inadequate; for here we have individuals with two X chromosomes, and yet (they) are not females. They are shifted out of the female class by the presence of an extra set of autosomes, and thereby the autosomes are proved to play a positive role in the production of sex. Since the intersexes differ from females by the assumption of certain male characters, this effect of the autosomes is due to an internal preponderance of 'male-tendency' genes.

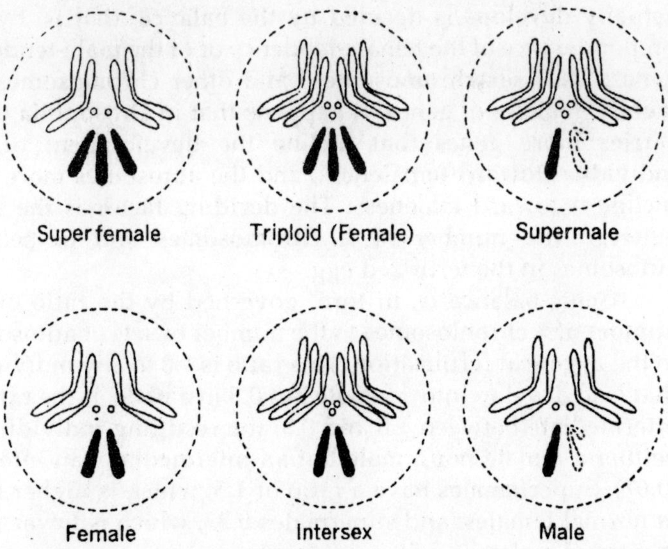

Fig. 7. Effect on sex of the balance between X chromosomes (solid) and autosomes (outlined) in *Drosophila melanogaster*.

We may now reformulate the sex relation as follows: Both sexes are due to the simultaneous action of two opposed sets of genes, one set tending to produce the characters called female and the other to produce the characters called male. These two sets of genes are not equally effective, for in the (chromosome) complement as a whole the female-tendency genes outweigh the male and the diploid (or triploid) form is female. The male-tendency genes in the autosomes are more numerous or more effective than those in the X chromosome, while the net effect of genes in the X chromosome is a tendency to the production of female characters. When in a diploid zygote the relative effectiveness of the female-tendency genes is lowered by the absence of one X, the male-tendency genes outweigh the female, and the result is the normal male. When the two sets of genes are acting in a ratio between these who extremes, as in the case in the ratio of 2X : 3 sets of autosomes (2X : 3A), the result is a sex intermediate—the intersex."

Genetic balance theory. The foregoing quotation from Bridges' classical work states the theory of sex determination by means of genic balance. Every individual has in his genotype both female and male potentialities. Which sex actually develops is decided by the balance, that is, by the preponderance of the female-tendency or of the male-tendency genes. The "sex chromosomes" and other chromosomes are merely vehicles of genes. It happens that in Drosophila the X carries more genes that incline the development of the individual toward femaleness, and the autosomes more that incline it toward maleness. The deciding factor is the ratio between the number of X chromosomes and of sets of autosomes in the fertilized egg.

Genic balance is, in turn, governed by the ratio of the number of X chromosomes to the number of sets of autosomes in the zygote at fertilisation. This ratio is 1.0 in any individual that is to develop into a female but 0.5 in a male. If the ratio is intermediate between 1.0 and 0.5, the resulting individual is neither a female nor a male but an intermediate, an intersex (0.67). Superfemales have a ratio of 1.5, which is higher than in normal females, and supermales 0.33, which is lower than in normal males.

Intersexes and Environment

A careful study of the intersexes in Drosophila has revealed a very interesting fact. As shown by Bridges, these intersexes all have two X chromosomes and three sets of autosomes. Yet the sexual characters of intersexes vary all the way from a condition resembling normal males (male-type intersexes) through various intermediate situations to a condition approaching normal females (female-type intersexes). If sex is determined by the ratio of X chromosomes to autosomes, what causes this wide variation among the intersexes?

Table 2
Sexual Types in *Drosophila Melanogaster*

Sex		*X chromosomes*	*Sets of autosomes (A)*	*Sex index (ratio X/A)*
Superfemale		3	2	1.5
Normal female	tetraploid	4	4	1.0
	triploid	3	3	1.0
	diploid	2	2	1.0
	haploid *	1	1	1.0
Intersex		2	3	0.67
Normal male		1	2	0.50
Supermale		1	3	0.33

* A whole individual of this type has not been obtained, but patches of tissue with one X and one of each autosome have been found in diploid flies. These patches show female characters.

Triploid females were allowed to produce offspring at different temperatures, and the sexual characteristics of the intersexes that appeared in their offspring were compared. The average sex type in the intersexes developed at high temperatures was shifted in the female direction, and at low temeratures in the male direction. High temperature has a feminizing and low temperature a masculinizing effect in Drosophila.

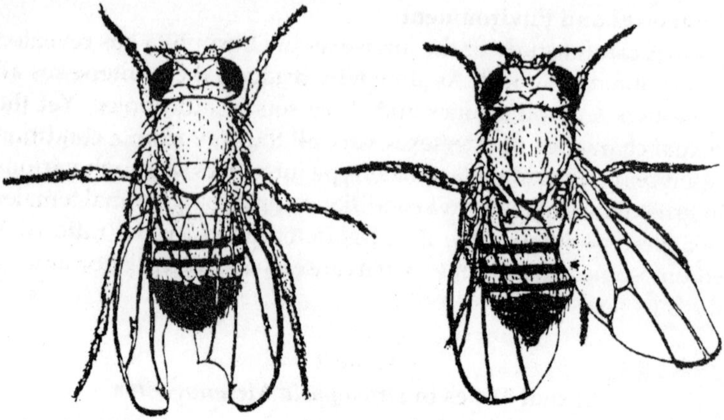

Fig. 8. Intersexes of Drosophila melanogaster with chromosome complements 2X: 3A. At left, male-type intersex; at right, female-type intersex. (*After Bridges*).

Some triploid females produce intersexes that, in a given environment, are on the average more feminine or more masculine than others. By selecting females that produce the most feminine or most masculine intersexes in several generations, it is possible to obtain triploid strains that produce only female-type or only male-type intersexes. All these intersexes have the same chromosomes, namely two X chromosomes and three sets of autosomes, but some of them have more feminizing and others more masculinizing genes. The chromosomal mechanism interacts with different genetic "environments" or "modifiers."

Intersexes in the Gypsy Moth

The classic work of Goldschmidt on diploid intersexes in the gypsy moth, *Lymantria dispar:* has made an important contribution to the understanding of the mechanisms of sex determination. This moth is remarkable for its striking *sexual dimorphism;* it occurs naturally in Europe, northern Africa, and northern Asia. The offspring of matings between females and males from any one geographic locality consist of females and males only, but if strains of European origin are crossed with those from Japan or if strains from northern and southern Japan are intercrossed, the progenies contain some intersexes. To account for these results, Goldschmidt supposes that sex in the gypsy moth is determined by the interaction of factors for maleness carried in the X and factors for femaleness carried in the Y or in the cytoplasm. These factors vary in "strength." In any one race, the strengths of these factors are so adjusted by prolonged natural selection that two doses of maleness (carried in two X chromosomes) dominate the femaleness and produce a normal male and one dose of maleness (one X chromosome) is dominated by the femaleness and gives rise to a normal female.

Table 3
Sexual Types Resulting from Crosses of European and Japanese Strains of the Gypsy Moth

Cross	*Parents*	*Sons*	*Daughters*
1.	European "weak" ♀ × Japanese "strong" ♂	Normal	Intersexual
2.	Japanese "strong" ♀ × European "weak"	Normal	Normal
3.	F_2 from cross 1	Normal	½ normal, ½ intersexual
4.	F_2 from cross 2	½ normal, ½ intersex	Normal

The sex factors are relatively "weak" in European and "strong" in Japanese strains. Accordingly, if a European female is crossed to a Japanese male, one Japanese X chromosome with its maleness is not entirely dominated by the relatively weak femaleness inherited from the European parent; the XY individuals that should develop into females become intersexes. In the gypsy moth, it is possible to distinguish intersexual females and intersexual males, that is, individuals which, according to their chromosomal constitution (XY or XX), should be females or males, respectively, but which become intersexes on account of the unbalance of the sex-determining factors in the different races.

According to Goldschmidt all intermediates between strong and weak sex factors can be found in different races. The "weakest" race of the gypsy moth lives on the northern island of Japan (Hokkaido), and the "strongest" race occurs at the northern extremity of the main Japanese island (Honshu). Crosses of these races result not in intersexuality but in sexual transformation (sex reversal); that is, some of the XY individuals develop into "transformed," or sex-reversed, males and XX individuals into "transformed" females. Thus the sex phenotype may be the opposite of the one chromosomally established.

As one proceeds southward from the northern tip of Honshu, "strong" races are met with less and less, until in southern Japan and Korea the strength of the sex factors is not very different from that encountered in typical European populations. Goldschmidt and his coworkers have also proposed a mechanism by which the sex factors influence the development of the sex phenotype. The essential assumption is that intersexes begin their development as females (or as males) and develop as such up to a certain critical point, the so-called "turning point," after which their development is of the opposite sexual type. The mixture of female and male traits found in intersexes is thus explained as due to the switching from the female developmental course to the male course, or vice versa, occurring earlier or later in development. A similar mechanism may operate also in the triploid intersexes in Drosophila, where the inersexes seem to start their development as males but complete it as females. The hypothesis of the turning point has, however, met with many objections and cannot be regarded as established.

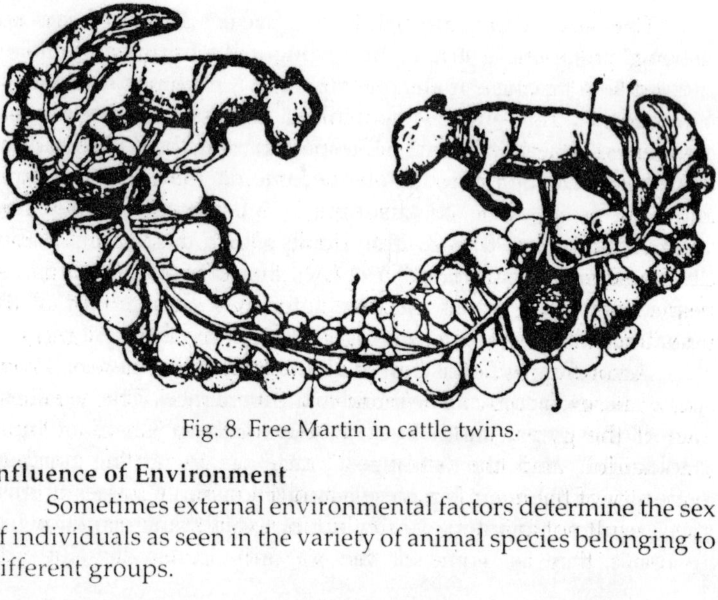

Fig. 8. Free Martin in cattle twins.

Influence of Environment

Sometimes external environmental factors determine the sex of individuals as seen in the variety of animal species belonging to different groups.

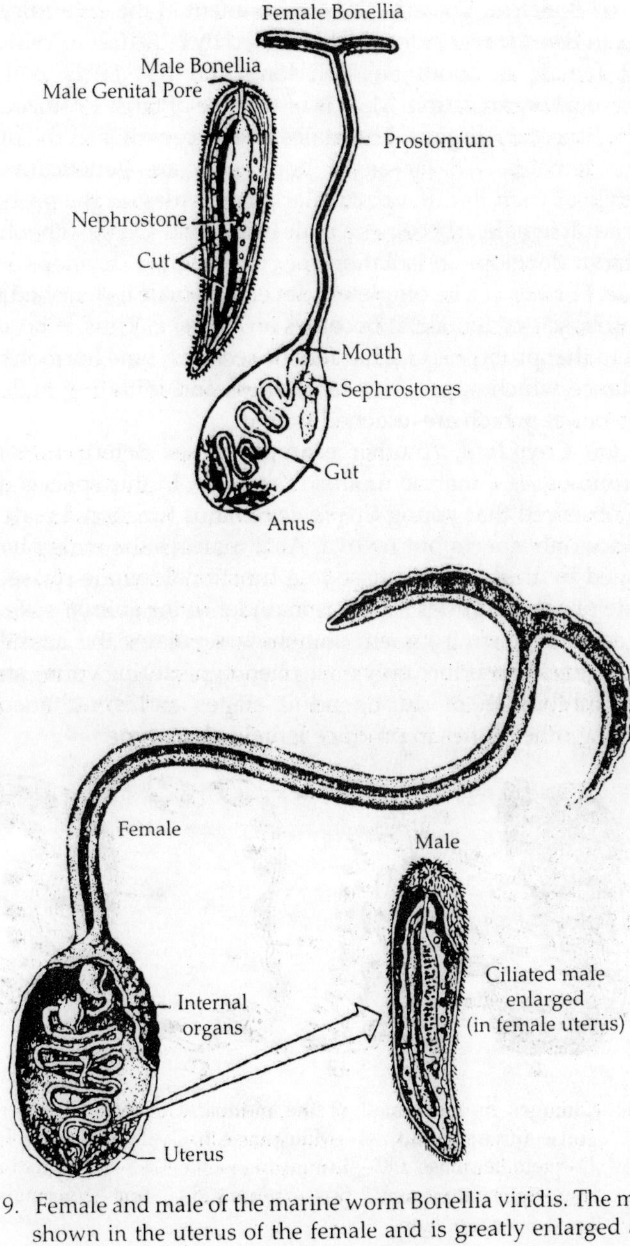

Fig. 9. Female and male of the marine worm Bonellia viridis. The male is shown in the uterus of the female and is greatly enlarged at the right to show details of internal structure.

(i) Bonellia. The effect of environment in the determination of sex in *Bonellia* was extensively studied by *F. Baltzer* in 1935. The adult female is about an inch long and has fairly complex anatomical organisation. Male is of the size of large Protozoa and has rudimentary organs. The males live as parasites in the uterus of the females. All larvae of Bon ellia are genetically and cytologically similar. If a particular larva settles on the proboscis of an adult female, it becomes a male individual. On the other hand, if a larva develops in isolation (i.e., in water) it develops into a female. Further, if a incompletely developed male is detached from the proboscis of female, it becomes an intersex. Thus, it becomes evident that proboscis of adult female secretes some hormone-like substance which suppresses femaleness and initiating maleness in the larvae which are attached with it.

(ii) Crepidula. Another example of sex determination by environment is a marine mollusc *Crepidula.* In this species it has been observed that young *Crepidula* remains functional male and produce only sperm but no ova. As it matures the male phase is changed by transitional phase to a functional female phase. The female phase continues for the remainder of the animal's life. It is suggested that environment somehow regulates the activity of certain genes to produce only male phenotype during young stages, while inhibits them during adult stages and simultaneously initiating other genes to produce female phenotype.

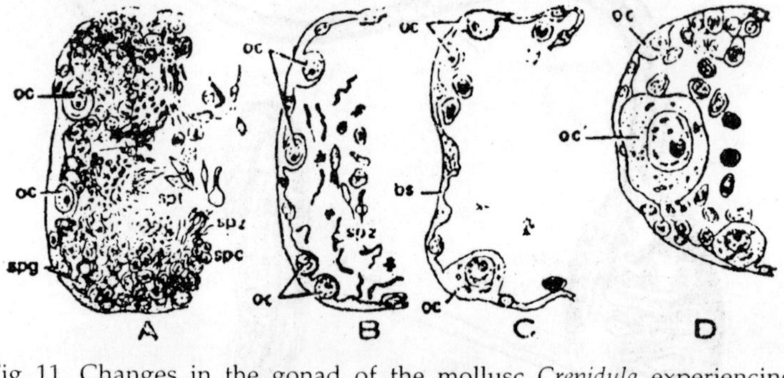

Fig. 11. Changes in the gonad of the mollusc *Crepidula* experiencing consecutive sexuality. A—Male phase, B, C—Transitional phases; D—Female phase. OC—Immature oocyte, OC'—Mature oocyte: Spg—Spermatogonia, Spt—Spermatids, Spa—Spermatozoa, Spc—Spermatocytes.

Ophryyotrocha. It is a segmented animal in which the male and female sexes are determined by the number of segments of a body has. If the number of segments is more than 20, the animal is female otherwise it is a male. By cutting the number of segments sex from female to male can easily be changed.

Dinophilus. It is a marine annelid. It produces two types of eggs—large and small. Larger eggs develop into females and the smaller one develop into males. Among adult worms, the females are much larger than the males. It appears that sex determination in *Dinophilus* depends upon the quantity of the cytoplasm in the egg cell which can be utilised for the development of the embryo.

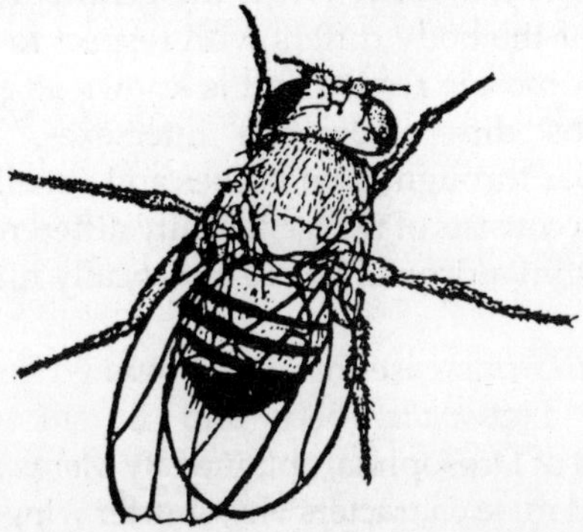

Fig. 12. A gynandromorph of *Drosophila melanogaster.* The left side of the body is predominantly female and the right side predominantly male.

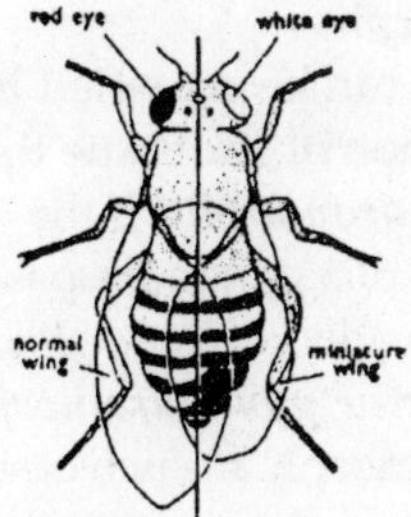

Fig. 13. Bilateral gynandromorphis in Drosophila.

Echiuroid Worm. Weisz (1971) describe that the sex determining factor in echiuroid worm in environmental CO_2.

Relatively low concentration of CO_2 is sea water causes free swimming larvae of this worm to develop as females. Added respiratory CO_2 produced by an adult female or with a larva already determined as a female then undetermined larvae develops as a male.

In Greek (*Gyne,* woman, *ander*man and *morphe,* form) the term gynandromorph refers to the appearance of both male and female characters at different halves of the body in the same organism. In insects, the secondary sexual characters are not influenced by the hormones secreted by the gonads and the differentiation occurs independently depending upon the genetic constitution of the cells in various parts of the body. When the genotype of the cells in different parts of the body differs with respect to the sex linked characters, a sex mosaic results and is known as *gynandromorph.* Gynandromorphs differ from the intersexes. The later are genetically similar throught their bodies and usually sterile, but a gynandromorph consists of two genetically different tissues. Some of the cells of a gynandromorph are genetically male, and others are female.

Gynandromorphic cases have been studied in several species of insects like *Drosophila,* bees and in silk worms. In a gynandromorph of Drosophila, obtained by *Morgan'* the right half of the fly showed male characters viz., shorter wings, black tipped abdomen and a sex comb of the first leg. The left side showed contrasinting female characters. The right eye was white and the left one red.

Types of Gynandromorphs

Gynandromorphs can be classified into the following types on the basis of the position of sex tissue the gynanders have:

(i) Bilateral gynandromorphs. If the line of division between male and female tissues of a gynander passes through the middle of the body the stage is called *bilateral gynandromorph.*

(ii) Anterio-posterior gynandromorphs. It is less common than bilateral gynandromorph. In such cases the anterior region of the body has the characters of one sex and posterior region has the character of other sex.

(iii) Sex piebalds. This kind of gynandromorph is of very rare occurrence. In this the body consists of female tissue having spots of male tissues scattered irregularly or the vice versa.

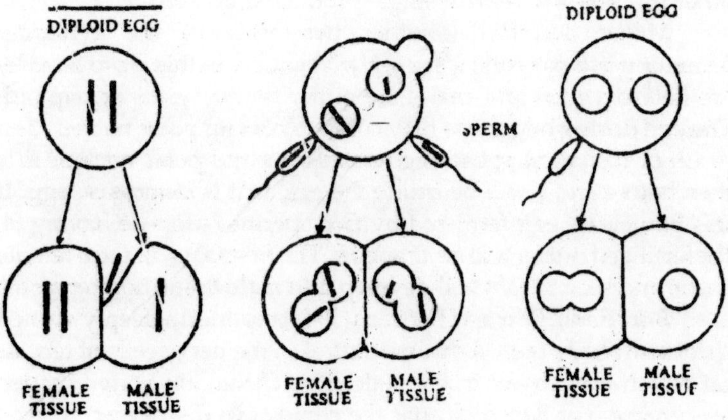

Fig. 14. Showing the origin of Gynandromorph.

Origin of Gynandromorphs

Three different hypothesis have been offered to explain the origin of gynandromorphs.

(i) By elimination of X-chromosome. Usually gynander starts its development with two X's. But in the later phase during cell division one X gets lost or disappeared in daughter cells. The two X chromosomes in the usual mitosis become divided into four X chromosomes (daughters). The one daughter receives only two X's and other daughter cells gets only one wile fourth X becomes lost due to abnormal cell division. The daughter cell receiving two X's forms the female tissue while other daughter cell having one X develops into male tissue. It occurs in *Drosophila*.

(ii) By retention of polar nucleus in the egg. In silk worms female is XY. During meiosis X and Y get separated, either of one (X or Y) going to egg and polocyte. But sometimes polar body nucleus remains in the egg alongwith nucleus. Thus, these eggs will be binucleate (XY). As a result of fertilisation two sperm cells (which contain X) fuse with X and Y chromosome separately giving rise to male (XX) and female (XY) tissues respectively. In *bees* also binucleate eggs are found (either due to fusion of polar body nucleus with the egg nucleus or parthenogenetically) producing gynanders. In such cases when fertilisation takes place one sperm cell fuses with the one nucleus of egg forming *female tissue* while

other nucleus will remain single producing *male tissue*.

Muller described another possibility of the gynander formation in a parasitic wasp—*Harbrobracon*. In this wasp females are heterozygous and males are either homozygous or haploid. The egg during oogenesis may come to contain polar nucleus as a result of fusion. Suppose egg nucleus is *a* and polar nucleus is *b;* then both *a* and *b* will be inside the egg as it is binuclease egg. If this binucleate egg fertilized by two sperms (suppose having *b*), the fertilized nuclei will be *ab* and *bb*. The first (*ab*) will form female tissue and second (*bb*) will develop into male tissue homozygous.

Barr Body. Barr and Bertram (1949) reported a deeply stained chromatin body (i.e., a chromocentre) in the nerve cells of female cat which was absent in the male. This chromatin body is called *sex chromatin* or *Barr body* after the name of its discoverer.

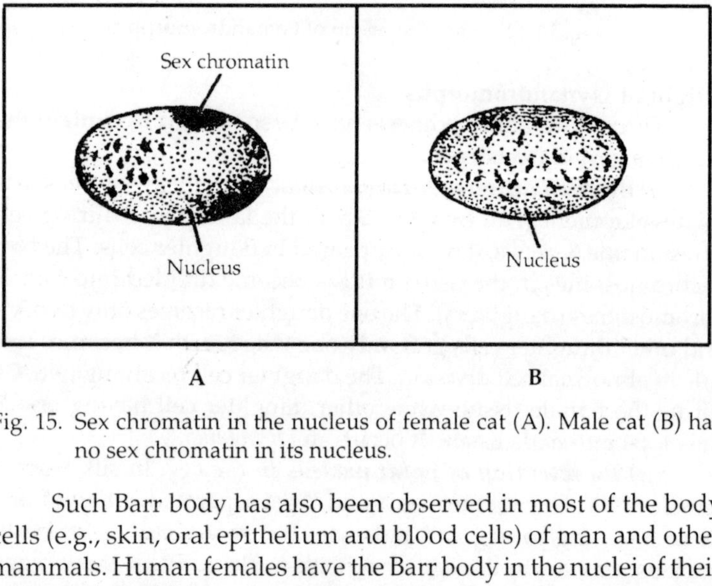

Fig. 15. Sex chromatin in the nucleus of female cat (A). Male cat (B) has no sex chromatin in its nucleus.

Such Barr body has also been observed in most of the body cells (e.g., skin, oral epithelium and blood cells) of man and other mammals. Human females have the Barr body in the nuclei of their body cells in higher proportion than males and are therefore, referred as *sex chromatin positive*. The human males are called *sex chromatin negative*.

Similar darkly stained sex chromatin has also been reported from the female neutrophils where it is called *drumstick body*. The relationship between sex chromatin and sex chromosomes has recently been elucidated (*Ohno, Kaplan* and *Kinosita* 1959). It has been demonstrated that the sex chromatin is derived only from one of the two X-chromosomes. The other X chromosomes behaves like an autosome and is not heteropycnotic at interphase. The number of corpuscles of sex chromatin at interphase is equal to nx-1. It means that there is one Barr body less than the number of X-chromosomes. Besides its genetical significance, Barr body has a wide field of medical application and offers the possibility of relating the origin of certain congenital diseases to X chromosome anomalies.

7

Sex Linked Inheritance

In autosomal inheritance the genes are found on the autosomes and their inheritance is unconditional whereas in sex linked inheritance the related genes are found on sex chromosomes specially on the X-chromosomes. It is the conditional type of inheritance. In this the sons resemble their mother and daughters resemble their father.

Sex-linked Inheritance

Thomas Hunt Morgan in 1990 in an early study of the fruit fly, *Drosophila Melanogaster* reported the normal eye colour of the fly which is brick red. In a laboratory population that had been maintained for many generations, Morgan found a single male with white eyes. In a mating of this male with red-eyed females, all the F_1 progeny had red eyes, as would be expected if the allele for white is recessive. In the F_2 produced by the mating of F_1 males and females there were 2459 red-eyed females, 1011 red-eyed males, and 782 white-eyed males. These numbers represent a rather poor fit to the expected 3:1 ratio of red-versus white-eyed phenotypes. Note also that all of the white-eyed flies were males, a surprising result. Furthermore, when a white-eyed male was crossed with his red-eyed female offspring, the progeny consisted of both red- and white-eyed males and red- and white-eyed females in approximately equal numbers, indicating that white eyes are not limited to males.

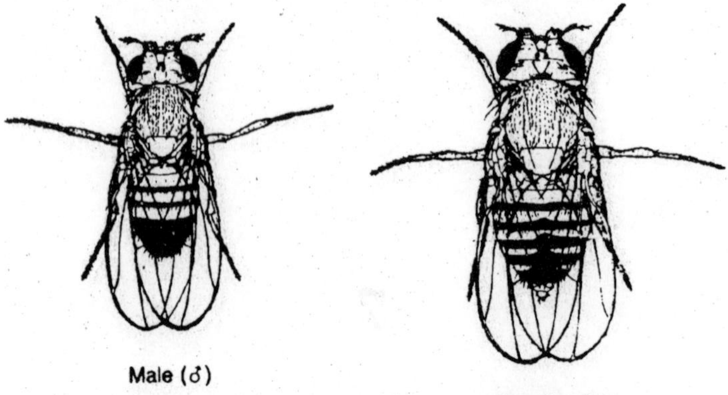

Fig. 1. Drawings of a male and female fruit fly Drosophila melanogaster. (Original drawings by Edith Wallace, artist for Thomas Hunt Morgan).

Another unexpected result was obtained when white-eyed females were mated with red-eyed males—the *reciprocal* of the cross with the original white-eyed male. In this cross the F_1 females had red eyes and the males again had white eyes. Morgan summarised the results of these crosses by stating that this trait (not all eye-colour variants) is *sex linked* in inheritance, meaning simply that it is associated with the inheritance of sex. It later became known that *Drosophila* females have two X-chromosomes and males have an unequal XY pair, from which the inheritance of the white-eye character could be explained by assuming that the alleles for red (w^+) and white (w) are located on the X chromosome and not present at all on the Y.

Morgan later discovered other genes that follow this same X-linked pattern of inheritance. Note that in a male only one copy of an X-linked gene is present, because the gene is not also on the Y chromosome. The term *hemizygous* is used to describe this condition.

Y-linked Genes

Despite major progress since the 1950s certain aspects of human sex determination are still not entirely understood. The X-Y basis of sex determination in humans and other mammals has been known since 1923; females have two X chromosomes (XX) and males have one X and one Y (XY). However, until the late 1950s, an open question remained unanswered: Which sex chromosome, the X or the Y, possesses the gene or genes responsible for governing the sexual destiny of a developing embryo? Some discoveries in the late 1950s and early 1960s seemed to point to the answer. Studies of individuals with abnormal numbers of sex chromosomes revealed that 45,X0 humans developed ovaries and 47,XXY developed testes. What did this indicate? Consistent with these findings, it was hypothesized that the Y chromosome possessed a genc that dictated development of a male. In 1987, David Page and his colleagues at the Whitehead Institute for Biomedical Research in Cambridge, Massachusetts, reported the result of studies that enabled them to pinpoint the location of a gene on the Y chromosome whose presence always resulted in the development of a male. The gene was called the *testis-determining factor (TDF),* and it is located in Yp11.2. The TDF gene was also found on the Y chromosomes of other species of mammals, including monkeys, dogs, horses and goats.

Recently, a specific gene has been tentatively identified as being the TDF gene in humans it is called *SRY,* sex-determining region of the Y chromosome. The following picture of human sex determination has now emerged. The development of male and female embryos proceeds identically until about six or seven weeks after fertilisation, when sexual differentiation occurs. At that time, a genetic signal initiates a cascade of biochemical events that leads to the development of a male. The genetic signal is hypothesized to be a protein encoded by *SRY* that regulates the expression of other genes. If the *SRY* gene is absent, the embryo develops into a female.

The story becomes more complicated because a nearly identical gene has also been found on the X chromosome. Its functional role, if any, in sex determination has not yet been defined. Thus in spite of the great excitement created by the discovery of the *SRY* gene the complete story of sex determination in human has not yet been deciphered, and research on this exciting question continues. A few other structural genes are located on the Y chromosome. Two or three Y genes encode specific proteins found on cell surfaces. Also, available evidence indicates that one or more genes affecting sperm production, stature (height). Tooth size, and hair on ear are located on the Y chromosome.

Are there any traits, other than maleness itself, that are determined by Y-linked genes? The inheritance of such traits could easily be detected because of their striking pedigree characteristics. Nevertheless, of the several thousand hereditary traits known, only one seems to follow the pattern of linked inheritance. This is a gene for the trait *hairy ears,* which refers to the growth of stiff hair an inch or longer on the outer rim of the ears of males beyond the age of about 20. The trait is most commonly found in men of India (incidence about 20 per cent) and Israel, but it also occurs in Caucasians, Japanese, and other ethnic groups. Y-linked inheritance of hairy ears is not entirely certain because the trait is difficult to study. Not only is it extremely variable in age of onset (in some men it is not expressed until 60 or 70), but the trait is also extremely variable in its expression. Nevertheless, pedigrees of hairy ears do suggest Y linkage because the trait occurs only in males and is transmitted only through males. Whether the hairy ear gene is truly Y linked or not, it is clear that the Y carries few genes.

X-Linked Genes

The X chromosome, in contrast to the Y, carries as many genes as would be found on an autosome of comparable size. The genes on the X (called *X-linked* genes or, less frequently, sex-linked genes) seem to be a random collection of genes that are no different in their functions from those found on the autosomes. Over 100 different loci are known to be X linked and nearly 100 others are strongly suspected to be. Among these loci are the *HGPRT* locus used routinely in somatic cell genetics, loci involved in colour perception and blood clotings, and a locus that, when mutateud, causes a degenerative disease of muscle known as *Duchenne-type muscular dystrophy.*

Pattern of Inheritance

The pattern of inheritance of X-linked genes is unique because males receive their X chromosome only from their mother and transmit it only to their daughters. This pattern is often called *crisscross* inheritance because an X chromosome can crisscross between the sexes in successive generations. A surname analogy for X-linked inheritance can be devised if we imagine a "strange savage nation" in which males have one surname but females have two. In this nation, a male's surname is one chosen at random from his mother, but female always receives her father's surname in addition to one randomly chosen from her mother. Thus, for example, a mating between Mr. Smith and Ms. Brown-Robinson would produce sons named Brown or Robinson with equal likelihood and daughters named Smith-Brown or Smith-Robinson with equal likelihood.

In this analogy, the names correspond to X-linked alleles. With just two alleles at an X-linked locus (call them *A* and *a*), there will be three possible genotypes in that, for rare autosomal recessive alleles, heterozygotes are much more frequent than recessive homozygotes. This same relationship is true in females with respect to rare X-linked recessive alleles. This is also valid for X-linked recessives if the numbers along the horizontal axis are interpreted as referring to the incidence of recessive homozygotes, among females and those along the vertical axis are interpreted as referring to the frequency of heterozygotes among females.

For traits due to X-linked recessive alleles, affected males occur much more frequently than affected females. This aspect of X-linked inheritance is illustrated in Figure. For a trait that affects 1 per cent of females (i.e., $\frac{1}{100}$), for example, the incidence of the trait in males is 10 per cent (i.e., $\frac{50}{500}$), which means that affected males are 10 times as frequent as affected females. To take another example, a rare trait that has an incidence of 1 per 1 million females will have an incidence in males of 1 per 1000 (i.e. $\frac{0.5}{500}$), so in this case affected females. Although the mathematical basis for the differences in incidence between the sexes will be discussed later, the reason for the differences can be seen intuitively: Whereas two copies of a recessive X-linked allele are required to produce an affected female, only one copy is required to produce an affected male; and an individual is less likely to inherit two copies of a rare allele than only one copy.

At this point it is convenient to discuss several examples of X-linked inheritance.

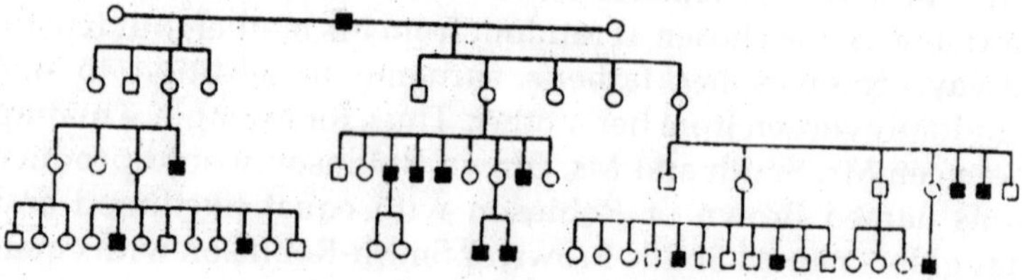

Fig. 2. Pedigree of halmophilia in a kinship from Scotland shown several typical features of the inheritance of rare X-linked mutations. Note that only males are affected. Because males transmit their X chromosome only to their daughters, and because the rare mutation is recessive, affected fathers have normal sons and carrier daughter. Moreover, half the sisters of affected males are expected to be carriers. Carrier females are expected to have 50 per cent affected sons.

Glucose 6-Phosphate Dehydrogenase (G6PD) Deficiency

One well-known gene on the X chromosome controls the production of an enzyme, *glucose 6-phosphate dehydrogenase* (G6PD), which is involved in carbohydrate metabolism and is important in maintaining the stability of red blood cells. Individuals who have abnormally low amounts of this enzymatic activity are prone to severe anaemia that occurs when many of their red blood cells cannot function normally and therefore break down and are destroyed. The anaemia can be provoked by a number of environmental triggers such as inhaling pollen of the broad bean *Vicia faba* or eating the bean raw, in which case the illness is known as *favism.* Such individuals are also sensitive to certain drugs such as naphthalene (used in mothballs), certain sulfa antibiotics (such as sulfanilamide), or the antimalarial drug primaquine. In the absence of the offending substances these individuals are completely normal, and they recover from the anaemia when the agents are eliminated. G6PD deficiency is found in high frequency in people of mediterranean extraction (10 to 20 per cent or more of males are affected) and among Asians (about 5 per cent of Chinese males are affected), and it occurs in about 10 per cent of black American males. As noted, the locus of G6PD is X linked, and well over 50 alleles coding variant forms of the enzyme are known. However, only a few of these alleles lead to a sufficiency defective form of G6PD to cause the drug sensitivity and anaemia associated with G6PD deficiency.

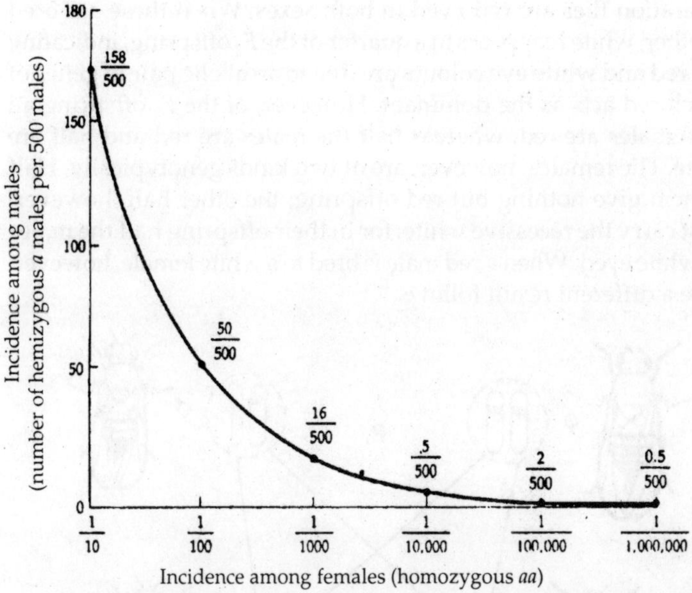

Fig. 3. Incidence of hemizygous males for an X-linked recessive (vertical axis) plotted against incidence of homozygous recessive females (horizontal axis). When an X-linked recessive is rare, there will be many times more affected males than affected females.

Sex Linked Inheritance in Drosophila

Events of crucial importance for the chromosome theory proved to be the discovery of sex-linked genes in Drosophila by Morgan in 1910 and coordinated genetical and cytological study of these cases by Morgan, Bridges, and others. In the course of breeding experiments with the normal wild type, which has red eyes, Morgan found one individual in which the eyes were white. This gave rise to a true-breeding strain of white-eyed flies. When he crossed this new variety with the wild, red-eyed type, the results from a cross of a white male with a red female were different from those obtained from the reciprocal cross of red male with white female. The results were found to depend on the sex of the parent by which the trait was introduced into the cross, whereas with other Mendelian characters, it makes no difference in either the F_1 or the F_2 whether a given character is brought in by the male or female parent.

From the cross of white-eyed male with red-female, the first-generation flies are red-eyed in both sexes. When these are bred together, white reappears in a quarter of the F_2 offspring, indicating that red and white eye colours are due to an allelic pair of genes of which red acts as the dominant. However, of the F_2 offspring all the females are red, whereas half the males are red and half are white. The females, however, are of two kinds genotypically. Half of them give nothing but red offspring; the other half, however, must carry the recessive white, for in their offspring half the males are white eyed. When a red male is bred to a white female, however, quite a different result follows.

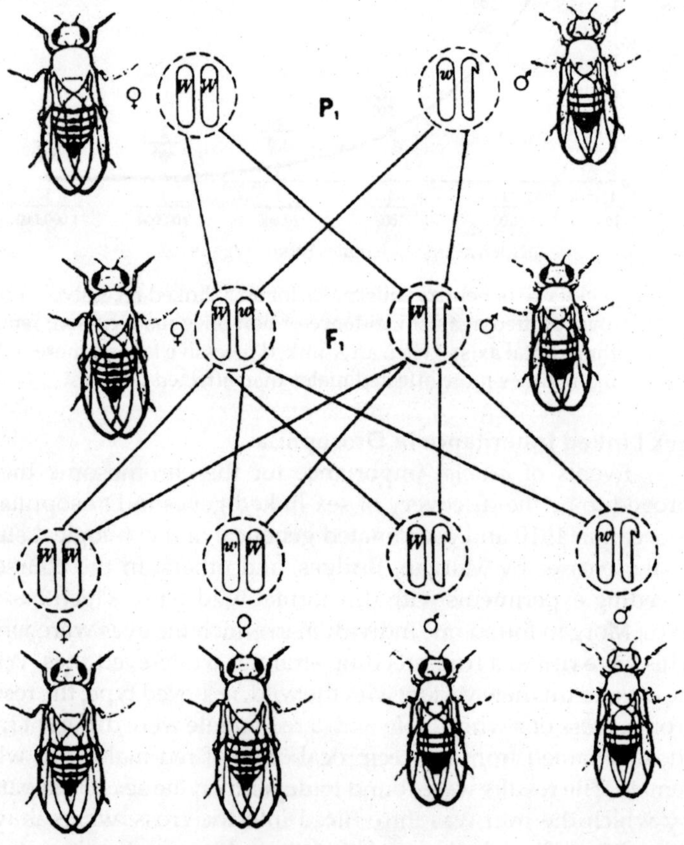

Fig. 4. Sex-linked inheritance in Drosophila. The cross of red-eyed female by white-eyed male. The course of sex chromosomes carrying the sex-linked gene W-w is traced from the parents to the F_2 Females at left, males at right.

Among their F_1 offspring, all the females are red-eyed and all the males are white-eyed. When these are bred together, their offspring (the F_2) consist of red-eyed and white-eyed individuals in equal numbers in both sexes. All the white-eyed flies are apparently pure, for no red-eyed flies appear in their offspring, and the red-eyed males bred to pure red females also produce only red-eyed descendants. The red-eyed F_1 females, however, must be heterozygous, for when bred to either white or red males, half their male offspring are always white-eyed. In Drosophila, sex-linked traits such as white eye colour follow a crisscross inheritance. The male transmits his sex-linked traits to his grandsons through his daughters, never to or through his sons. The trait thus seems to alternate or cross from one sex to the other in its passage from generation to generation. This, of course, is the mode of transmission followed by the X chromosome, as can be seen in the diagrams. Only the daughters get an X chromosome from the father, whereas both sons and daughters receive an X chromosome from the mother. In explaining the inheritance of white eye colour in Drosophila, it was assumed that the gene for white eyes is located in the X chromosome and that the Y chromosome carries no normal allele for white.

The white-eyed female crossed with a red-male transmits a gene for white to each offspring and an X chromosome to each offspring. The daughters receive also an X chromosome from the father carrying the dominant allele of white and hence are red-eyed. The sons, however, receive a Y chromosome from the father and hence no allele of white, and they are thus white-eyed. The gene for white follows exactly the transmission of the X chromosome. Accordingly, Morgan concluded that this gene is actually carried in the X chromosome. Sutton and Boveri put forth the general hypothesis that Mendelian genes are borne in chromosomes, but Morgan was the first to associate a specific gene with a specific chromosome visible in microscopic preparations. At least 140 other genes in *Drosophila melanogaster* follow the same mode of inheritance as white, which indicates that these genes are also carried in the X chromosome.

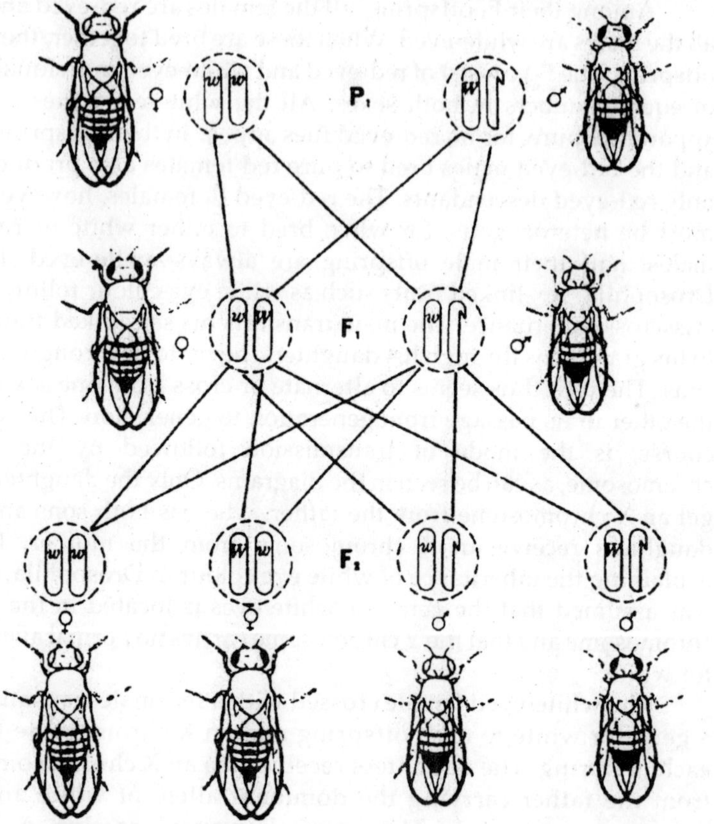

Fig. 5. Sex-linked inheritance in Drosophila. The cross of white-eyed female by red-eyed male.

Sex-Linked Inheritance in Poultry

In poultry several genes are located in the X chromosome, and their inheritance follows the same general scheme as outlined for Drosophila, cats, and man. However, the females is the heterogametic sex. The hen has but one X chromosome and no Y. This is known as an XO condition. She produces two kinds of gametes, one with an X chromosome and one without. The male is the homogametic XX, producing only one kind of gamete, that with the X chromosome. This can be illustrated best by the gene for barred feathers, found particularly in the Barred Plymouth Rock breed. Instead of solid colouring, the feathers are banded with "bars" of black on a white background. This gives the characteristic barred effect.

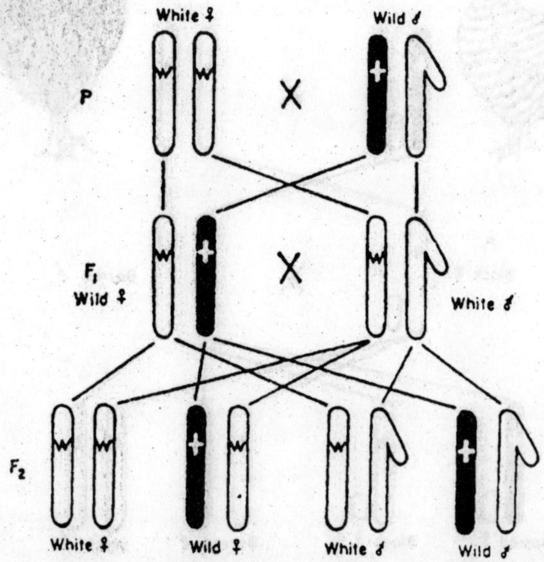

Fig. 6. Sex-linked inheritance in Drosophila.

The gene for characters showing sex linkage are located in the X chromosome of the XY pair concerned with the determination of sex. In Drosophila and most mammals, including man, the females is the homogametic sex (XX), while the male is heterogametic (XY). The Y chromosome is composed mainly of heterochromatin with few genes. It contains nothing to mask the expression of any character whose determiner (gene) is in the X chromosome. Therefore, one X chromosome of XY individuals alone determines the inheritance of gene located in it. XY individuals are hemizygous, and transmit only the X chromosome to their daughters. In Drosophila and mammals, males transmit an X chromosome to their daughters, never to their sons. In poultry, females transmit an X chromosome to their sons, never to a daughters.

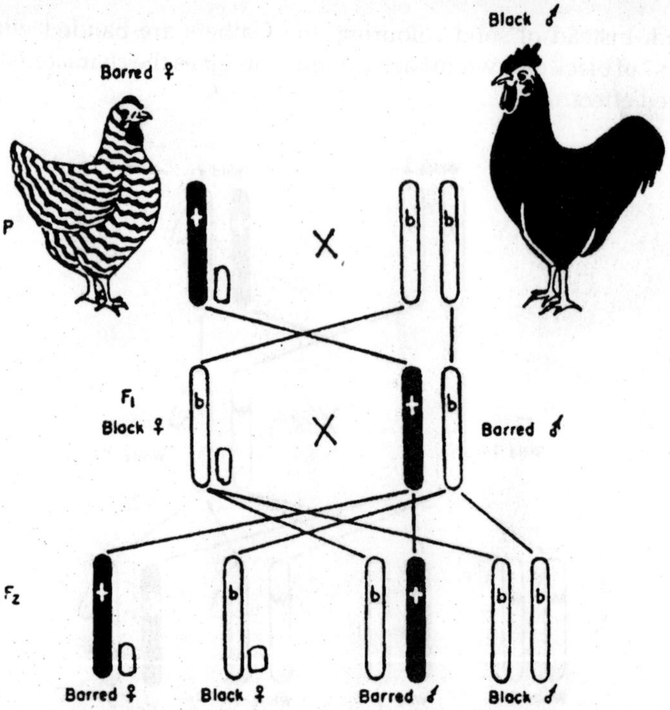

Fig. 7. Sex-linked inheritance in poultry. A mating of a barred female to a black male is illustrated. In poultry the female has but one X chromosome with no Y (an X-O condition). The male has two X-chromosomes.

Sex-linked Inheritance in Cats

Tortoise-shell cats are those with blotches of black-yellow and white in their coats. It was also stated that such cats were always females. This phenotype is produced in the heterozygous condition of the gene determining black or yellow coat colour. If we designate black colour as *B/B*, or preferably +/+, and the yellow colour as *b/b*, then the F_1 females would be of the constitution +/*b*, and would be neither black nor yellow, but tortoise-shell. Note that the females are the same regardless of the way the cross is made. This is because *neither* black nor yellow is dominant, but the heterozygous is distinguishable from both homozygotes. The male, being hemizygous can be either yellow or black, but not both. It should be noted that occasional tortoise-shell male cats have been observed. They have been sterile, however, most likely because of an additional X chromosome.

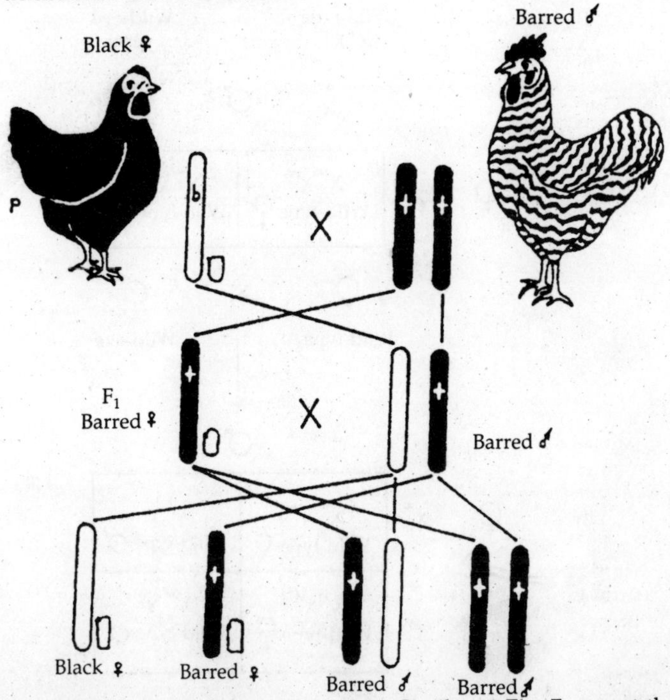

Fig. 8. Sex-linked inheritance in poultry. Similar to Fig. 7, except that a black female is mated to a barred male.

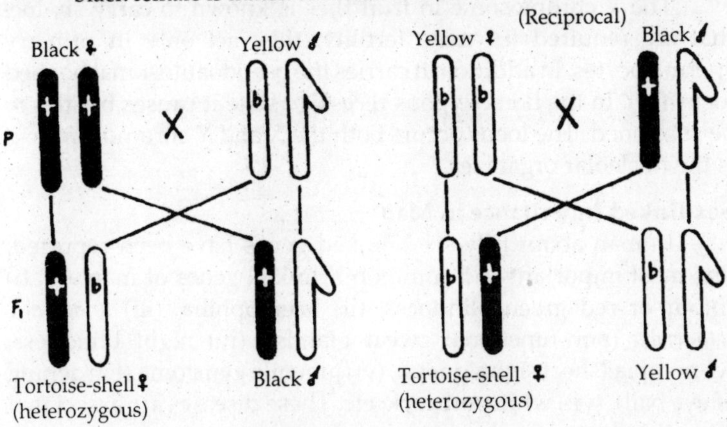

Fig. 9. Sex-linked inheritance of coat colour in cats.

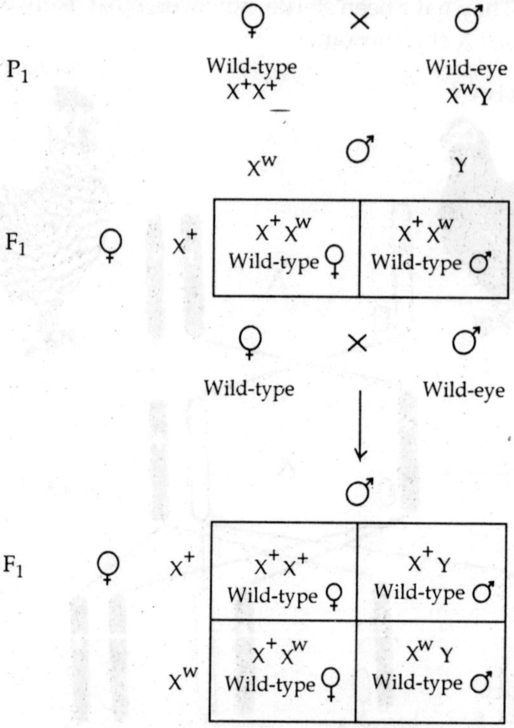

Fig. 10. Crosses of figures redrawn to include the sex chromosomes.

The Y chromosome in fruit flies is known to carry six loci that are required for male fertility; they act only in primary spermatocytes. In addition, it carries the pseudoautosomal bobbed locus (*bb*). In the homozygous recessive state it causes bristles to be shortened. The locus occurs both the X and Y chromosomes: it is the nucleolar organizer.

Sex-linked Inheritance in Man

In man about fifty-six X-linked genes have been reported. The most important and common X-linked genes of man are: (i) colour or red green blindness, (ii) haemophilia, (iii) *anhidrotic ectoderma* (non-functional sweat glands), (iv) night blindness, (v) myopia (short sightedness), (vi) juvenile glaucoma (hardening of eye ball), (vii) white fore lock, etc. These diseases are associated with X-linked recessive genes and are most common in man.

Certain X-linked diseases such as defective tooth enamel which results in early wearing down of the teeth, however, is caused by dominant genes.

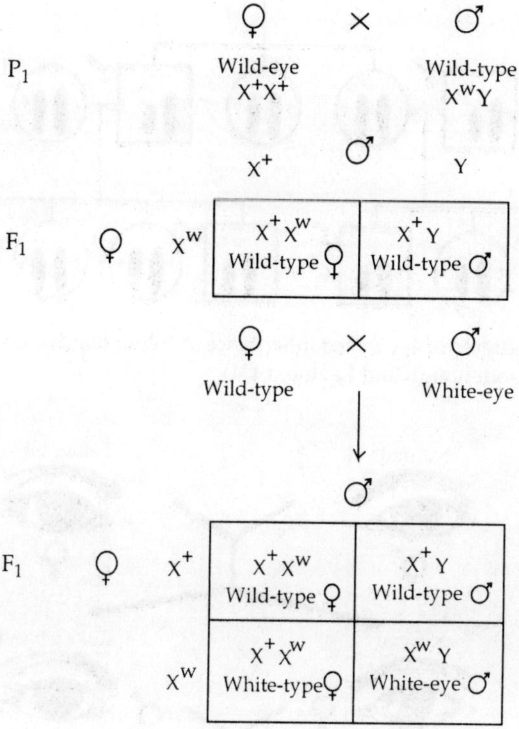

Fig. 11. Reciprocal cross to that in figure 9.

1. Colour blindness. The retina of man eyes contains colour sensitive cells which are necessary for the distinction of red and green colours. These colour sensitive retinal cells are formed and controlled by certain genes which remain located in X chromosomes. The man having the recessive genes of above mentioned genes remain unable to distinct the red and green colour so are known as *colour blind* or *red green blind* man. This X-linked recessive gene is able to express itself in male due to absence of its any allelic gene in the Y chromosome.

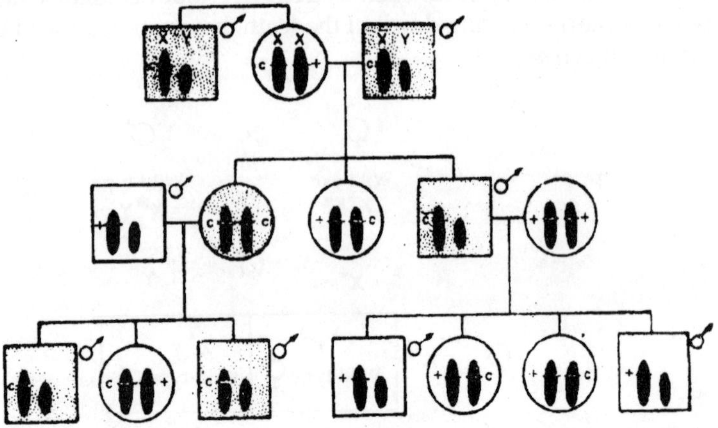

Fig. 12. Pedigree of sex-linked inheritance of colour blindness in man (after Goodenough and Levine, 1974).

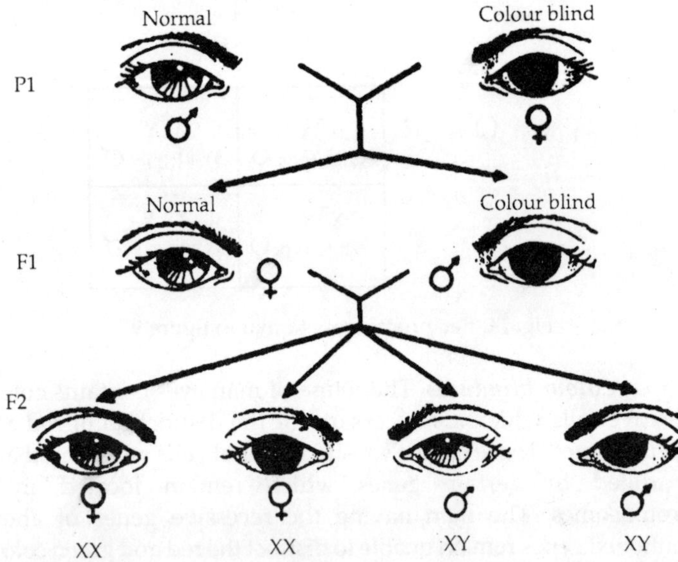

Fig. 13. Cross between a red-green colour blind female (XX) and a normal male (XY) showing X-linked inheritance in man.

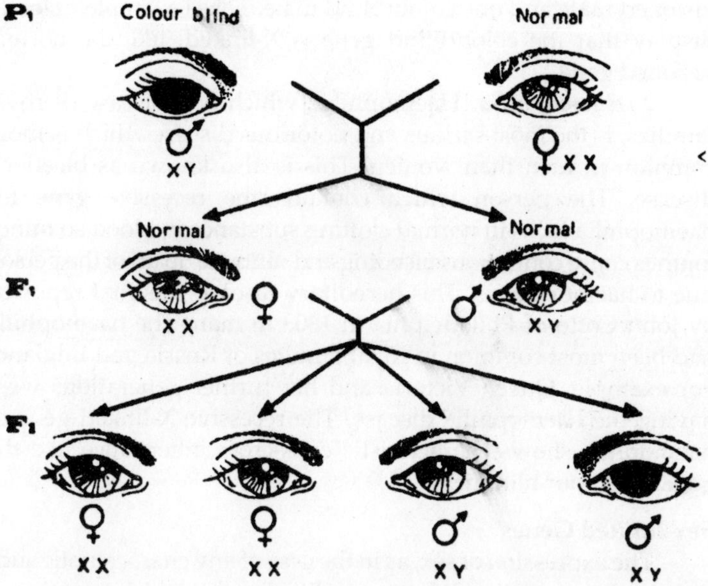

Fig. 14. Cross between a red-green colour male and normal female showing X-linked inheritance in man.

The inheritance of colour blindness can be studied in following two types of marriages:

(i) Colourblind woman and normal visioned man. If a woman (XX) is colour blind and she happens to marry a normal visioned male (XY), then all F_1 sons will be colour blind and daughters will be normal visioned. Because male receives one X-linked recessive gene for colour blindness from colour blind mother. The daughter receives a X-linked dominant gene for normal vision from father and one X-linked recessive gene for colour blindness from the mother. The F_1 brother and sisters, if inbred or married they will produce in F_2 a colourblind homozygous daughter, a normal visioned heterozygous daughter, a normal vision hemizygous son and a hemizygous colourblind son.

(ii) Colourblind man and normal woman. When a colourblind man marries with a normal visioned woman, then they will produce normal visioned daughters and sons in F_1. The marriage between a F_1 normal visioned woman and normal visioned male, will produce in F_2 two normal visioned females, one normal visioned male and one colourblind male. These examples, clearly display that the colourblind gene is X-linked, like the normal visioned gene.

2. ***Haemophilia.*** Haemophilia, which is a disease of royal families, is the most serious and notorious disease which is more common in men than women. This is also known as bleeder's disease. The person which contains the recessive gene for haemophilia lacks in normal clotting substance in blood so minor injuries cause continuous bleeding and ultimate death of the person due to haemorrhages. This hereditary disease was first reported by John Cotto of Philadelphia in 1803 in man. The haemophilia had been most common in royal families of Russia and England. For example, Queen Victoria and her further generations were having the haemophilia disease. The recessive X-linked gene of haemophila shows characteristic criss-cross inheritance like the gene for colour blindness.

Sex-limited Genes

The expression of sex, as in the case of any characteristic such as eye colour and blood clotting ability, involves the interaction of many genes with one another and with the environment. While the genes for the switch to maleness causing the onset of testis development are holandric (Y-linked), they are not the only ones which play a role in the development of the male reproductive system or of other masculine features. Other genes, including those for normal differentiation of testes and other parts of the male reproductive system, may be on autosomes or even on the X. Similarly, while the X chromosome carries genes necessary for the normal differentiation and expression of many female traits, other genes located on autosomes also exert an influence on female development.

The X is a large chromosome which carries the loci for many genes which influence characteristics unrelated to sex. All the observations are compatible with the concept that expression of sex and other human characteristics depends upon the actions of an assortment of genes found throughout all the chromosomes. And these activities may be influenced by the environment in which they take place. Once these points are appreciated, it comes as no surprise that certain traits which are commonly considered male or female are determined by genes carried by both sexes. For example, beard growth is considered masculine, since it normally takes place in males.

Genes determining beard features are not confined to the Y chromosome, even though the expression of the characteristic is normally confined to males. That genes on autosomes are involved is indicated by the fact that beard features found in a father and his son may be very different. Genes which a male receives from his mother can evidently affect his beard growth. Therefore, a female must carry these genes which typically do not express themselves in her. Such genes whose expression is normally limited to just one of the sexes are called *sex-limited genes*. Their limited expression results from the nature of the hormonal environment. In the internal environment of members of both sexes, both male and female hormones are present. It is the proper balance of these hormones which determines the normal expression of certain genes.

Genes for beard growth will act only when male hormone is present above a certain level. In some pathological conditions, there may be an elevation of male hormone in a female. The adrenal cortex secretes many types of hormones, among them a class of androgens or male hormones. The presence of a tumor can cause the level of androgens to rise in a female; as a consequence, beard growth may occur. The role of the internal environment is also clearly seen in the expression of breast development, which usually takes place only in the female. A male as well as a female, however, carries genes for this female characteristic. Therefore, a girl may show breast features which are due to genes transmitted to her by her father. She would receive a corresponding set from her mother as well, but some of these may not be expressed due to dominance or to other kinds of genic interactions.

Genes for breast development depend upon a certain level of estrogen for their expression; so they are sex-limited, since the necessary level is not usually reached in the male. Klinefelter males, often show some breast development, whereas the Turner female shows little. The atypical expression of sex-limited genes such as those for the breast and the beard characteristics undoubtedly ensures from upsets in the female: male hormonal balance caused by unbalance in the chromosomal constitution. Since hormones of both sexes are present in every person, any factor which increases the amount of male hormone in the female can bring about virilization; likewise, an abnormal increase in the proportion of female to male hormones in the male may cause feminization.

The role of hormones in triggering the expression off sex-limited genes is evident in certain cases of precocious puberty, where sexual maturation takes place at a rate faster than the normal. Development of a deep voice, which depends on male hormone, as well as other traits associated with masculine maturity, may be displayed by a boy who has not reached the age typically associated with the expression of certain secondary sex characteristics. The cause of such precocious expression may be a tumor of certain regions of the testis or of the adrenal cortex which brings about increased levels of male hormone. From cases such as these, it becomes clear that hormones may influence the time of expression of a gene. Any factor which alters the normal hormone balance at a given time can thus have a pronounced effect on the rate and development of a characteristic. Precocious puberty may also occur in a girl and may stem from pathological conditions in which estrogen levels become elevated at a premature age.

Sex-influenced Genes

The *sex-influenced genes,* genes which are expressed in both sexes but which behave as dominants in one sex and recessives in the other. Sex-influenced genes may be carried on any chromosome, but the majority of them is not sex-linked. The genes involved in pattern baldness appear to be sex-influenced. In the typical pattern baldness, onset of hair loss occurs about the age of 30 in the males. During the next 10 years or so, hair is lost in a typical fashion, usually starting from the top of the head and proceeding outward in all directions. Studies of human pedigrees are compatible with the concept that a pair of alleles is involved: *B* for baldness and *b* for non-bald. The genes are autosomal and are inherited in a typical Mendelian fashion. However, the gene *B* for baldness is expressed as a dominant in males and as a recessive in females. Its allele *b* for the non-bald trait acts as a dominant in females and a recessive in males. Therefore, for a woman to become bald, she must have the homozygous constitution *BB*. For a male to retain his hair, his genotype must be *bb*. The expression of the heterozygous state, *Bb*, is influenced by the sex of the individual, producing a bald phenotype in a male and a non-bald one in a female. There is thus a decided effect on the dominance of each of these 2 alleles which is caused by the sex of the person.

As in the case of sex-limited genes, the hormonal environment appears to be responsible for the difference. That this is the case is illustrated by those conditions where hormone therapy has been prescribed. A male with a sparse head of hair, when given female hormones for clinical reasons, may start to grow hair on the top of his head. Such a person may be genotype *Bb*. Under the usual circumstances, his male hormone level permits *B* to be expressed as a dominant. When given astrogens, the increased amount of female hormone permits the gene *b* to be expressed. However, this hair growth is often accompanied by breast enlargement, since the sex-limited genes for breast development, carried by the male but normally suppressed, will be called to expression. In a comparable manner, a woman taking male hormone may suddenly begin to lose her hair. This could happen to a woman of genotype *Bb* in whom the increased amount of male hormone will allow the *B* gene for baldness to mask the expression of its allele, *b*. Moreover, a male who is castrated and loses his prime source of male hormone will express the genes for pattern baldness in the same way as would a female; and so a male castrate who is genotype *Bb* would tend to retain his hair. Hair loss in a female of the genetic constitution *BB* will not be as pronounced as in a male of the same genotype. This is understandable in view of the fact that any normal woman would have a higher level of female hormone than of male hormone, and this would influence the degree of expression of the *B* gene, even when it is present in the homozygous condition.

As might be expected, other genes play a role in baldness, and at least one pair is suspected to influence the time of onset of baldness. Young men may go bald if they carry genes for early hair loss, but these genes appear to be sex-limited. In women, any genes for early loss of hair cannot express themselves; consequently a female with normal hormonal balance will not lose her hair at any early age, even though she may have the appropriate genetic constitution. It must be pointed out that not all cases of baldness are genetic. Hair may be lost for a variety of strictly environmental reasons: high fever which can destroy hair follicles; certain chemicals and microorganisms which can cause hair to fall. Some persons may thus be bald even though they have the genotype conducive to good hair growth. These persons would be phenocopies whose phenotype resembles that resulting from a specific genotype.

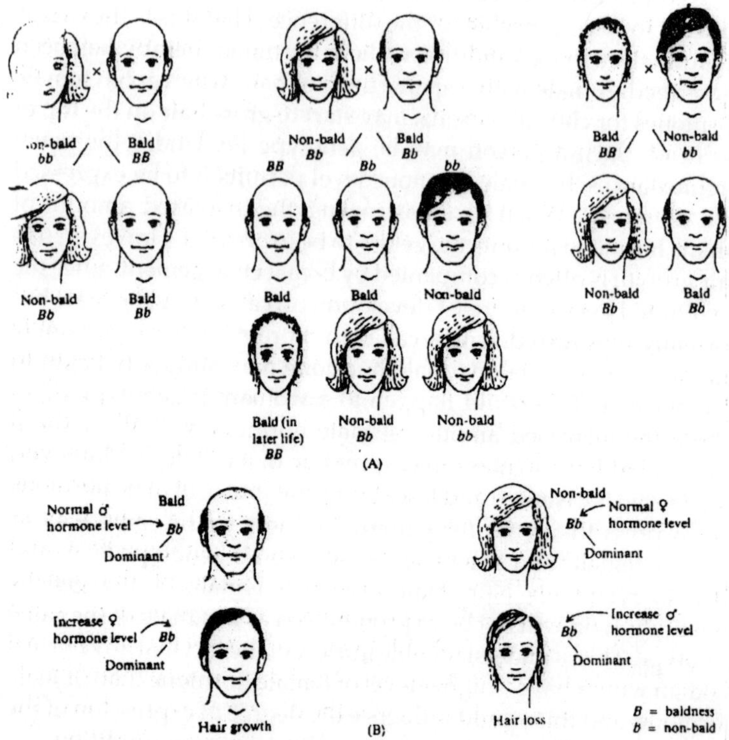

Fig. 15. Sex-influence genes. (A) The inheritance of pattern baldness is no different from that of any other autosomal trait. The expression of the genes involved in baldness, however, is influenced by the hormonal environment so that the gene B behaves as a dominant in males and as a recessive in females. Note that a woman (as in the cross on the right) will not show the extreme degree of baldness as will a male, even when she does experience hair loss. Moreover, the onset of hair loss will not occur until later in life. (B) The interaction between the genetic and the hormonal environment becomes very apparent when males must take female hormones or when females are prescribed male hormones. If two such persons are heterozygotes, the increased female hormone level will trigger gene *b* to expression in the male who is bald, and hair growth will occur. On the other hand, the increased male hormone in the woman under treatment will call gene *B* to expression, and hair loss will occur.

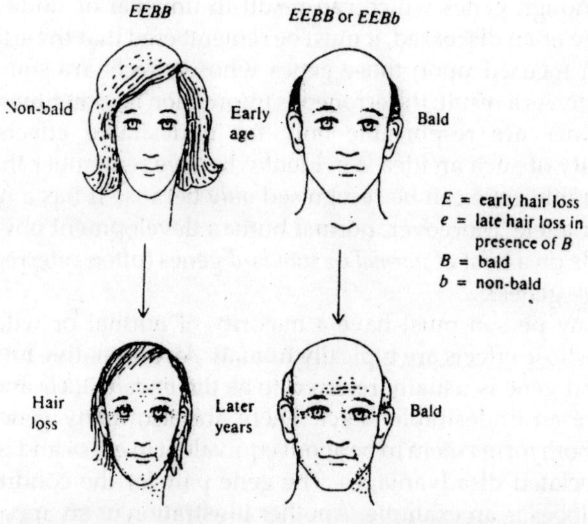

Fig. 16. Sex-limited genes. Hair growth is under the influence of more than one pair of genes. At least one pair of them (*E, e*) appears to influence the *time* at which hair loss takes place. This pair of genes seems to be sex-limited. The gene for early loss of hair, *E*, cannot express itself in young women. It is limited in its expression to males. Therefore, even if a woman has the genetic constitution, *BB,* for hair loss, she will express these genes only in later life.

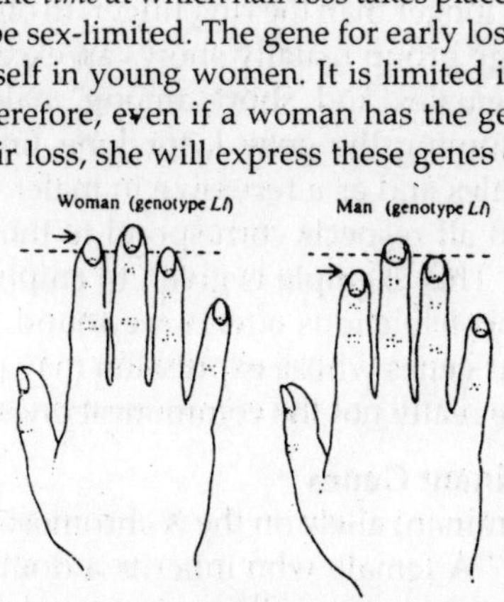

Fig. 17. Length of index finger. This is an example of an insignificant characteristic which is based on the expression of a pair of sex-influenced genes. *L,* (long index finger) and/short index finger). When the ring finger is placed on a line, it will be found that the index fingers of some persons extend above the line (left), whereas those of others fall below it (right). The gene *L* behaves as a dominant in females and as a recessive in males. Therefore, the genotype *LI* is expressed as long finger in a woman and as short finger in man.

Though genes which can result in unusual or undesirable traits are often discussed, it must be remembered that the attention is often focused upon those genes whose effects are somewhat dramatic. As a result, the erroneous impression is sometimes given that genes are responsible only for undesirable effects. The absurdity of such an idea is evident when we remember that any undesirable gene can be recognised *only* because it has a normal form or allele. Moreover, normal human development obviously depends on a host of *normal* or *standard* genes (often referred to as *wild-type* genes).

Any person must have a majority of normal or wild-type genes whose effects are typically human. An alternative form of a standard gene is usually referred to as the *mutant allele* and may produce an undesirable effect. There are also many gene pairs which both forms seem to be about equivalent in effect and to have no associated disadvantages. The gene pair for the condition of the earlobes is an example. Another illustration of an apparently innocuous characteristic in humans is the length of the index finger. A pair of sex-influenced genes seems to play a role. In some persons, the index finger is longer than the ring finger. In others, it is shorter. A survey of a large group usually shows an excess of long index fingers among females and short among males. This can be explained by assuming the gene *L* for long fingers to act as a dominant in females and as a recessive in males. The inheritance patterns would in all respects correspond to those discussed for pattern baldness. This example is given to emphasize that genes with relatively inconspicuous effects are found in every human genotype and that genes whose expression may prove drastic for the individual are really not the commonest ones.

Sex-linked Dominant Genes

How is adominant allele on the X chromosome expressed in each of the sexes? A female who inherits a dominant sex-linked allele has the associated trait or illness, but a male who inherits the allele is usually more severely affected because he has no other corresponding allele. The children of a normal man and a woman with a dominant, disease-causing gene on the X chromosome face the following risks:

male, severely affected = ¼
male, not affected = ¼

female, affected has phenotype) = ¼
female, not affected = ¼

An example of a sex-linked dominant condition is incontinentia pigmenti. The name derives from the major sign in females with the disorder-swirls of pigment in the skin that resemble swirls of paint or marble cake. The skin pigment melanin leaks down into the deeper skin leayers, causing the odd pigmentation pattern. The pattern tends to lessen by adulthood. A newborn girl with incontinentia pigmenti has yellow, pus-filled vesicles on her limbs that come and go over the first few weeks. Then the lesions become warty, and are eventually replaced by brown splotches that may remain for life. Other symptoms include hair loss, visual problems missing or peg-shaped teeth, and seizures. Males with the condition are so severely affected that they die in the uterus. This is consistent with the fact that women with the disorder have a high rate of miscarriage, about 25 per cent. Another of the rare sex-linked dominant conditions is vitamin D resistant rickets. There is also a very rare sex-linked dominant condition that causes extremely dense body hair.

XYY and Criminality

Another sex-chromosomal abnormality associated with few physical manifestations is the XYY condition—technically 47, XYY. Affected males have 47 chromosomes; they have the full chromosome complement of normal males plus a second Y chromosome. XYY males tend to be tall, averaging over 6 ft. Beyond this, distinctive physical symptoms are practically nonexistent, so it is questionable whether the term *syndrome* really applies to this case because there is no group of characteristic symptoms. Among new born males, the incidence of the XYY condition is thought to be approximately 1 in 950.

The behaviour of XYY males have been a matter of concern and controversy since 1965, when a Scottish study uncovered an incredibly high frequency of the XYY condition among imprisoned men convicted of violent crimes. More than 50 similar studies have now been carried out in prisons in Europe and the United States, and it has held true that the frequency of XYY men confined for crimes of violence is about 10 times greater than it is among a sample of new born males. Many of these studies have concentrated on tall men; because XYY males tend to be tall, this will automatically create a bias, and the findings must therefore be interpreted with caution. Nevertheless, the higher than expected frequency of XYY males among tall criminals has suggested to some that XYY males may undergo impaired psychosocial development, perhaps caused by their height itself, which may produce social stresses tending to evoke violent, aggressive behaviour, or perhaps caused by some subtle and invisible behavioural effect of the extra Y.

Table 1
Height, Criminality and Intellectual Function Among XY, XYY and XXY Danish Males

	Number of males	*Height, cm*	*Army intelligence test score*	*Educational index*
Average of all XY males	4096	187.1	43.7	1.6
XY males with no criminal record	3715	187.1	44.5	1.6
XY males with criminal record	381 (9.3%)	186.7	35.5	0.7
Average of all XYY males	12	190.8	29.7	0.6
XYY males with no	7	191.3	31.6	0.7
XYY males with criminal record	5 (41.7%)	190.2	27.0	0.5
Average of all XXY males	16	189.8	28.4	0.8

One Danish study of criminality among XYY males is particularly important and revealing because it is free of many of the biases in other studies. The Danish research involved an examination of all available records on almost all men 28 to 32 years old whose height was 184 cm (72.4 in) or more and whose birthplace was Copenhagen. A total of 4139 sex-chromosomal determinations were made; among these there were 12 XYY (incidence 1 per 345), 16 XXY (incidence 1 per 259—the XXY sex-chromosomal constitution will be discussed in a few pages), and 13 individuals with other types of chromosomal abnormality. The information examined included criminal records and types of crime, socioeconomic status of the men's parents, score on a Danish army intelligence test, and an index of educational level. Parental socioeconomic status was lower among men with criminal convictions than among those without convictions, but no differences were found among XY, XYY, and XXY males. Thus, lower socioeconomic status is associated with criminality but not with sex-chromosomal constitution.

Among XYY males, the rate of criminal convictions was 41.7 per cent (5 out of 12), which is significantly greater than the 9.3 per cent among XY males. (The term **significant** as applied to statistical tests means that an observed difference is greater than could plausibly be attributed to chance.) Among XXY males, the rate of criminal convictions (3 of 16) was somewhat greater than among XY males, but the difference is sufficiently small that it could be due to chance. On the average, XYY males are taller than tall XY males about 3.7 cm (1.5 in), and XXY males are taller than tall XY males by about 2.7 cm (1.1 in). Tallness alone is not related to criminality, however, because men with criminal records are somewhat shorter than their noncriminal counterparts.

One of the most important findings of the Danish study is that XYY and XXY males have a significantly lower score on the Army Intelligence test and a significantly lower index of educational level than do XY males. For each sex-chromosomal class, moreover, men with criminal records are substantially poorer in both measures of intellectual function than men without criminal records. Of course, this does not necessarily imply that men with lower levels of intellectual function actually commit more crimes; they may merely be more likely to get caught. In any case, the Danish data seem to imply that the high rate of criminality among XYY males is due to their moderately impaired mental function.

Some comments are in order about the nature of the crimes committed by the XYY males. Two of the XYY males were habitual criminals. One had more than 50 convictions for petty larceny, auto theft, burglary, embezzlement, and so on, but no record of violent or aggressive behaviour. The other habitual criminal, who was also mentally retarded and had spent most of his life in institutions for the mentally retarded, had convictions for arson, theft, burglary and embezzlement, but, once again, no convictions for aggressive crimes against people. The other three XYY males were petty criminals. Two had each been convicted twice for petty theft. The other had turned in a false report about a traffic accident and had started a small fire. Overall, *the crimes of the Danish XYY males were not crimes of violence and aggression; they were mainly crimes involving property.*

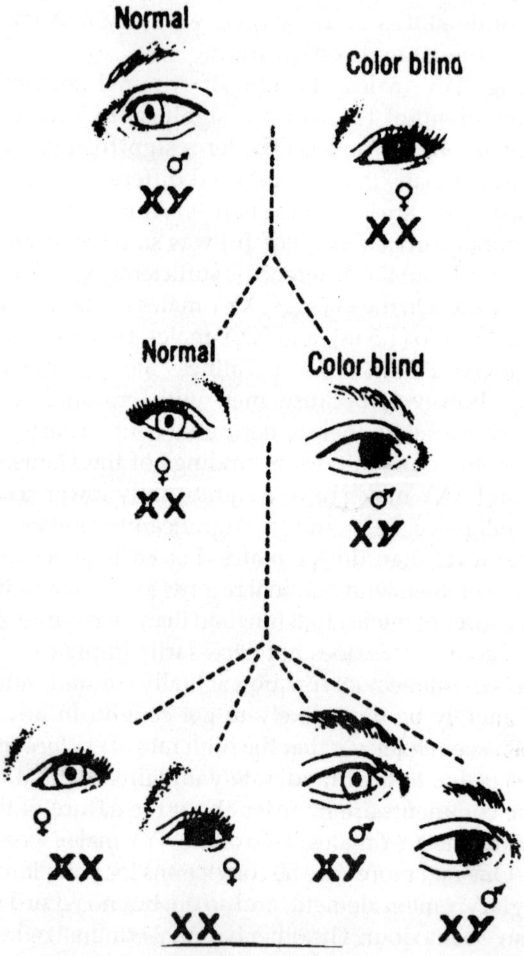

Fig. 18. The inheritance of red-green colour blindness in man. On the right, the result of marriage of a colour-blind man and a normal woman. The defect is transmitted only through the daughters and appears in half their sons, being carried in one of the X chromosomes.

The Danish study illustrates the danger of preliminary scientific reports being sensationalised by the popular press. At about the time of the original Scottish discovery of XYY males among exceptionally violent criminals, a pathological killer sneaked into a dormitory in Chicago and brutally murdered several student nurses. A newspaper claimed that the man had been "born to kill" because he was XYY. This turned out to be untrue—the Chicago killer was, in fact, XY–but the false report was widely disseminated and believed. Several proposals were made for the mass screening of newborns for the detection of XYY males to provide them with special education and intensive psychological guidance to counteract their supposed "killer instinct." It is good that the screening proposals were never carried out because, in the atmosphere of the time, they could have been harmful by creating a self-fulfilling prophecy. That is to say, if physicians and family are convinced that XYY males (or any other category of people) have severe, aggressive behavioural problems, then they may interact with the individuals in such a manner as to elicit the very behaviour they expect, thereby fulfilling their own prophecy. In light of the Danish study, the rate of criminality is related more to level of intellectual function than to sex chromosomal constitution.

Primary Nondisjunction

Morgan's identification of the microscopically visible X chromosome as the seat of the genes for white eyes and for other sex-linked traits in Drosophila was still based on an analogy: if one supposes that these genes are borne in the X chromosome, then the observed mode of their inheritance necessarily follows. Bridges, a student of Morgan, completed the proof. He showed that, taking the assumption that genes are carried in the chromosomes as a working hypothesis, one can predict how the chromosomes of certain flies will appear under the microscope. And, conversely, knowing the chromosomes, one can predict the behaviour of certain traits of the flies in breeding experiments.

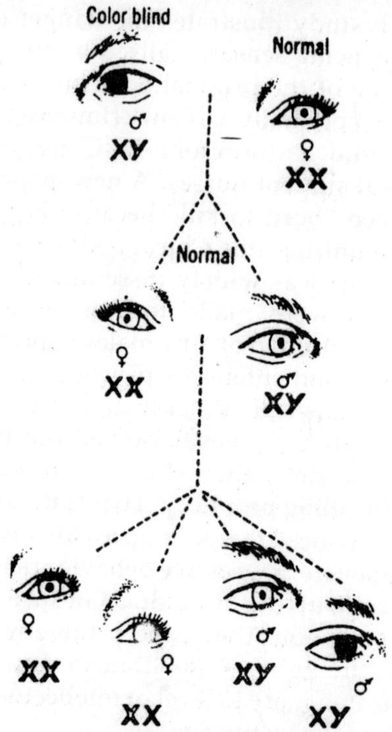

Fig. 19. Colour-blind individuals and chromosomes carrying the gene for this character are shown in black. On the left, results of marriage of a colour-blind woman and a normal man where the defect is transmitted to all the sons and to both grandsons and granddaughters.

When Bridges made the crucial experiments, the results came out as predicted. He published an account of his work in 1916 in a classical paper characteristically entitled "Nondisjunction as Proof of the Chromosome Theory of Heredity". As we have seen white-eyed Drosophila females crossed with red-eyed males produce red-eyed daughters and white-eyed sons in the F_1 generation. However, there are some exceptions to this rule. About one fly in 2,000 to 3,000 F_1 offspring has the unexpected eye colours, red in males and white in females. Does the appearance of these exceptional flies, even though they are rare, contradict Morgan's assumption that the gene differentiating the white-eyed flies from the red-eyed is transmitted in the X chromosome?

Bridges decided that there is no such contradiction. His reasoning was as follows. The white-eyed exceptional females must carry two X chromosomes, since their sex is female; and they must have inherited both their X chromosomes from their mothers, since the colour of their eyes is white. Similarly, the red-eyed exceptional males must have one X chromosome. Since they are males, and this X chromosome must have come from their fathers, since their eyes are red. How could such a distribution of the chromosomes arise? Bridges supposed that the two X chromosomes present in a female may occasionally fail to disjoin at the meiotic division as they should normally do.

The *primary nondisjunction* leads, then, to production of eggs with two X chromosomes and of eggs with no X chromosome. Such exceptional gametes, eggs of white-eyed females, will be fertilized by normal spermatozoa of red-eyed males. This fertilisation will give four types of zygotes, which will give rise to (1) white-eyed females, (2) red-eyed males, and (3) and (4) individuals with three and with no X chromosomes which can be supposed to be inviable. The validity of Bridges's supposition can be submitted to a rigorous test, for it leads us to expect that the exceptional white-eyed females have a peculiarity that no female was known to possess until then, namely, they must carry not only two X chromosomes but also a Y chromosome, which is normally present only in males. Bridges verified this prediction by examination of the chromosomes of the exceptional white-eyed females under the microscope. They actually have two X chromosomes and a Y chromosome (XXY females).

The exceptional red-eyed males should have no Y chromosome, which normal males of Drosophila always have. This prediction was also verified by cytological examination.

Secondary Nondisjunction

The exceptional males without the Y chromosome are sterile, although they are quite normal males in appearance and in behaviour. The white-eyed XXY females are, however, normal and fertile. Bridges crossed them to normal red-eyed males and observed in their offspring the *secondary nondisjunction*. In their progeny, about 96 per cent of the daughters have red eyes and about 4 per cent have white eyes. Among the sons, about 96 per cent are white-eyed and 4 per cent are red-eyed. At meiosis in

XXY females, the distribution of the X and Y chromosomes occurs. Four kinds of eggs are formed, namely, (1) eggs with a single X chromosome, (2) eggs with an X and a Y chromosome, (3) eggs with two X chromosomes, and (4) eggs with a Y chromosome.

Fertilised by the spermatozoa of a normal red-eyed male, these eggs should produce the eight types of zygotes. These eight types of zygotes will not occur with equal frequency since, at meiosis in XXY females, the two X chromosomes disjoin in about 92 per cent of the oöcytes but pass together either in the polar body or in the egg nucleus in about 8 per cent of the oöcytes. Examination of figure shows several ways of testing the validity of this seemingly very complicated working hypothesis. In the first place, all the white-eyed females and some of the red-eyed ones must carry not only two X chromosomes but also a Y chromosome. The red-eyed males, in contrast to those arising from primary nondisjunction, must have a Y chromosome and accordingly ought to be fertile. And finally, some of the white-eyed males must have a hitherto unknown set of chromosomes, namely, one X chromosome and two Y chromosomes. Bridges not only made these predictions but verified them by cytological examination of the proper classes of flies.

Attached X Chromosomes in Drosophila

In 1922 L.V. Morgan found another exceptional case of inheritance of sex-linked traits in Drosophila, and analyzed it along lines similar to those followed earlier by Bridges. This analysis proved the same principle, but in an even more diagrammatic way than that used in Bridges's original experiments.

Yellow body colour in *Drosophila melanogaster* is, like the white eye colour described above, due to a sex-linked recessive gene. Yellow females crossed with normal, brownish-gray males ordinarily give in the F_1 all normal daughters and yellow sons and equal numbers of yellow and gray in both sexes in the F_2. Normal females crossed to yellow males give all normal offspring in the F_1 and in the F_2 all normal females and a segregation 1 yellow: 1 normal among the males.

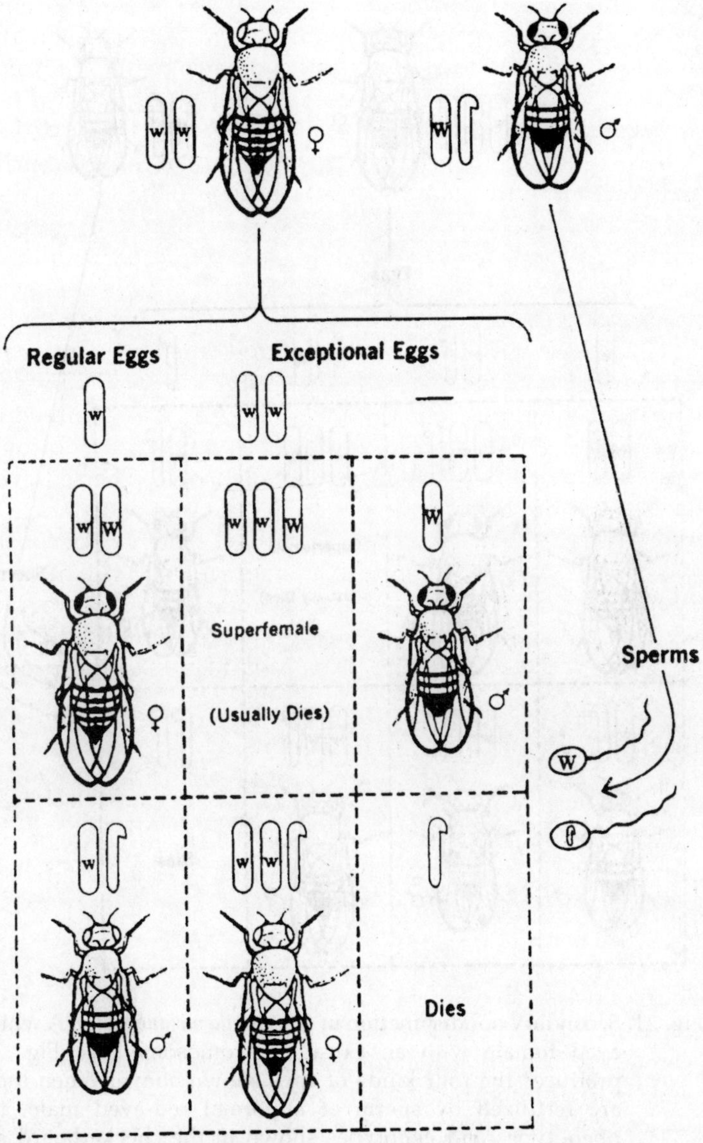

Fig. 20. Nondisjunction of the X chromosomes in *Drosophila melanogaster* in a cross of white-eyed female by red-eyed male.

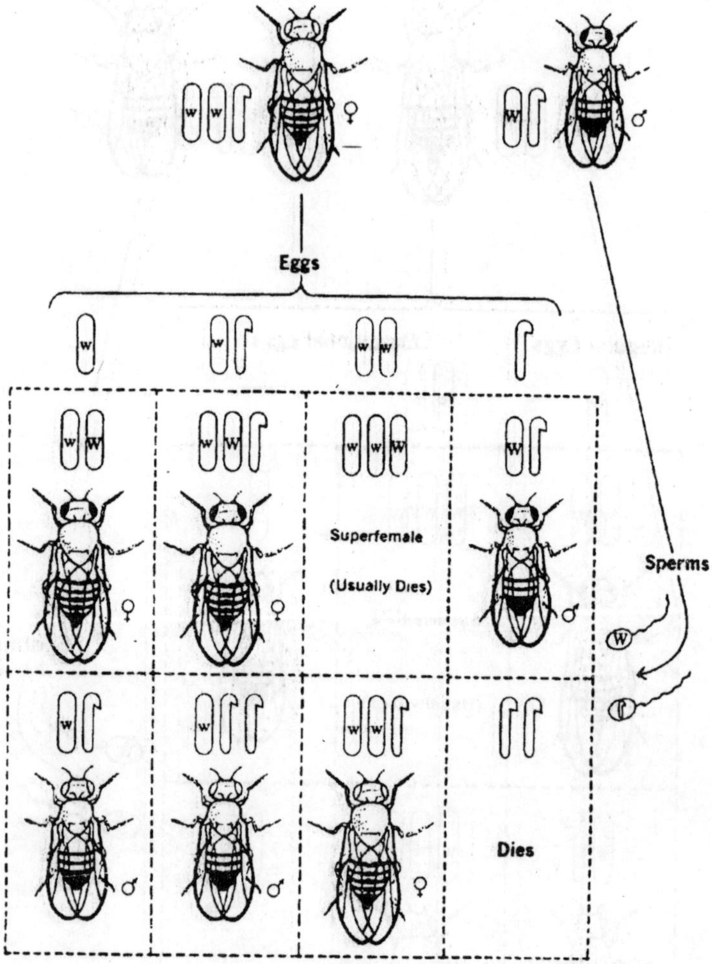

Fig. 21. Secondary nondisjunction in *Drosophila melanogaster*. A white-eyed female with an extra Y chromosome (see Fig. 20) produces the four kinds of eggs shown above. When these are fertilized by sperm of a normal red-eyed male, the phenotypes and genotypes shown in the checkerboard are produced.

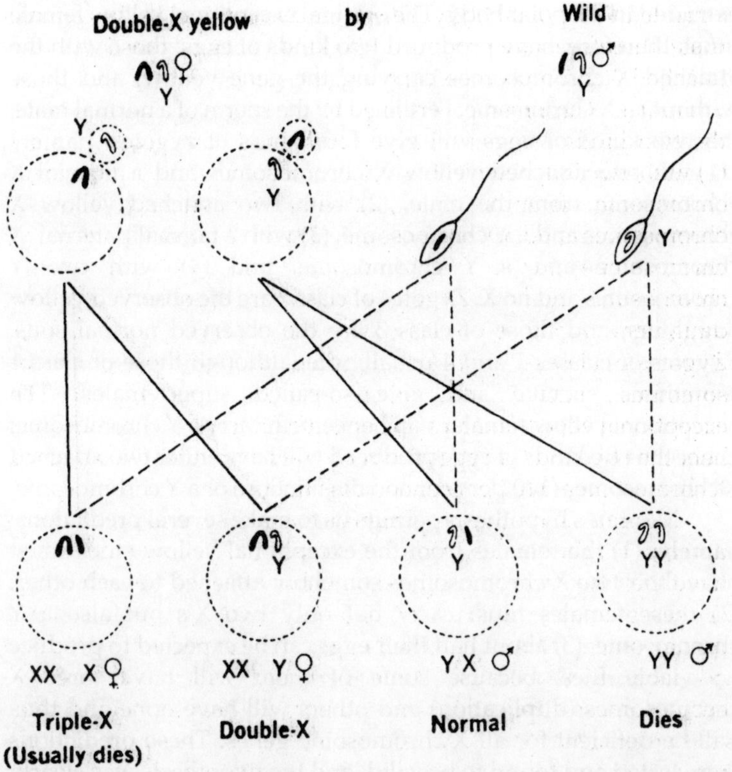

Fig. 22. Inheritance of "attached X" chromosomes in Drosophila. The X chromosomes of the mother each carry the gene yellow, and these are transmitted together. The only surviving sons are those that receive the Y chromosome from the mother and the X from the father.

L.V. Morgan found, however, an exceptional yellow female which, when crossed to normal males, produced only yellow daughters and normal sons. These yellow daughters repeated the exceptional performance of their mother; the sons of the original exceptional mother were sterile, but in subsequent generations the normal sons behaved like the usual normal flies. What is the explanation for this exceptional behaviour? Let us suppose that in the exceptional yellow female the X chromosomes become permanently attached to each other so that meiotic disjunction is no longer possible and both X's either remain in the egg or else are extruded in the polar body. The original exceptional yellow female must, therefore, have produced two kinds of eggs: those with the attached X chromosomes carrying the gene yellow, and those without an X chromsome. Fertilized by the sperm of a normal male, the two kinds of eggs will give four kinds of zygotes, namely (1) with two attached yellow X chromosomes and a normal X chromosome from the male, (2) with two attached yellow X chromosomes and a Y chromosome, (3) with a normal paternal X chromosome and a Y chromosome, and (4) with two Y chromosomes and no X. Zygotes of class 2 are the observed yellow daughters, and those of class 3 are the observed normal sons. Zygotes of classes 1 and 4 usually die, although those of class 1 sometimes survive and give so-called superfemales. The exceptional yellow females will henceforth carry a Y chromosome; hence the two kinds of eggs produced will have either two attached X chromosomes (100 per cent non disjunction) or a Y chromosome.

Morgan's hypothesis permits us to make several predictions: namely, (1) the females from the exceptional yellow stock must have their two X chromosomes somehow attached to each other; (2) these females must carry not only two X's but also a Y chromosome; (3) about half their eggs can be expected to produce no viable flies, because some of them will have three X chromosomes (duplication) and others will have none and thus will be deficient for all X chromosome genes. These predictions were tested and found to be valid, and the hypothesis was clearly verified. Females with attached X chromosomes have been found also in a related species, *Drosophila simulans*.

8

Non-Chromosomal Inheritance

The transmission of some inherited traits in eukaryotes does not follow a Mendelian pattern of transmission. These traits are usually determined by DNA molecules carried in cytoplasmic organelles (specifically, mitochondria and chloroplasts), by viruses, bacteria-like particles, or by other components in the cytoplasm. No criterion is universally applicable to distinguish nonchromosomal or cytoplasmic from nuclear inheritance. In higher organisms transmission of a trait through only one parent, typically the mother, usually indicates cytoplasmic inheritance. The cellular basis for this uniparental inheritance is that the major contribution to the cytoplasm of the zygote is from the egg, because sperm contain little cytoplasm. A genetic effect of uniparental inheritance is that the progeny produced by reciprocal crosses have different phenotypes. That is, the progeny from a mutant mother and normal father will be mutant, whereas a cross between a normal mother and a mutant father will yield normal progeny. This phenomenon, called a *maternal effect,* is usually indicative of extranuclear inheritance; however, an example given in the last section of the chapter will show that a maternal-inheritance pattern can also be produced by nuclear genes.

Evidences in Favour of Non-chromosomal or Cytoplasmic Inheritances

1. ***Maternal influence.*** When reciprocal crosses for a certain character are found to exhibit different results, it has been noticed that progeny exhibits the characteristics of their female parent. If chromosomal difference can be ruled out, maternal inheritance usually implies transmission through cytoplasm.

2. ***Reciprocal crosses.*** In cases of chromosomal heredity, the results of reciprocal crosses are ordinarily identical but in the cases of cytoplasmic inheritance the results of reciprocal crosses are different.

3. Infection-like transmission. When a heritable phenotype is transmitted without nuclear transmission, it suggests that some particles from the cytoplasm of the parent/parents have been transmitted to the offsprings.

4. Non-segregation and non-Mendelian segregation. Segregation is characteristic of Mendelian and chromosomal inheritance. Failure to show segregation may indicate extra-chromosomal heredity. Even if segregation occurs but in a fashion inconsistent with the segregation of chromosomes, the results may suggest the possibility of nonchromosomal inheritance.

5. Indifference to nuclear substitution. When a particular genotype exhibiting a specific character is replaced by a nucleus having alternative genotype, it does not change the phenotype and suggests the possibility of cytoplasmic inheritance.

Chloroplast Variegation in Four o'clock Plants

A well-understood example of cytoplasmic inheritance in higher plants is leaf variegation in the *albomaculata* strain of the four o'clock plant *Mirabilis jalapa.* Variegation refers to the appearance of yellowish-white (chlorophyll-free) regions in the leaves and stems of plants. The plants have some branches that are completely green, some completely white, and others variegated. All branches produce flowers, and these flowers can be used in crosses.

Nine distinct crosses can be made, in which pollen from flowers borne on white, green or variegated branches is used to pollinate flowers from white, green, or variegated branches. Seeds are collected and planted, and the phenotypes from the crosses are examined. Two significant observations are the following:

1. Reciprocal crosses yield different results. For example, the cross green female × white male yields green plants, whereas the cross white female × green male yields white plants (which usually die shortly after germination because of lack of chlorophyll and hence photosynthetic activity).
2. The phenotype of the female plants in each case determines the phenotype of the progeny (columns 1 and 3 in the table have the same entries). The male makes no contribution to the phenotype of the progeny.

This phenomenon is an example of *maternal inheritance.* When flowers borne on either the green or variegated progeny plants are used in subsequent crosses, the patterns of maternal inheritance that occur are identical to those observed in the original crosses.

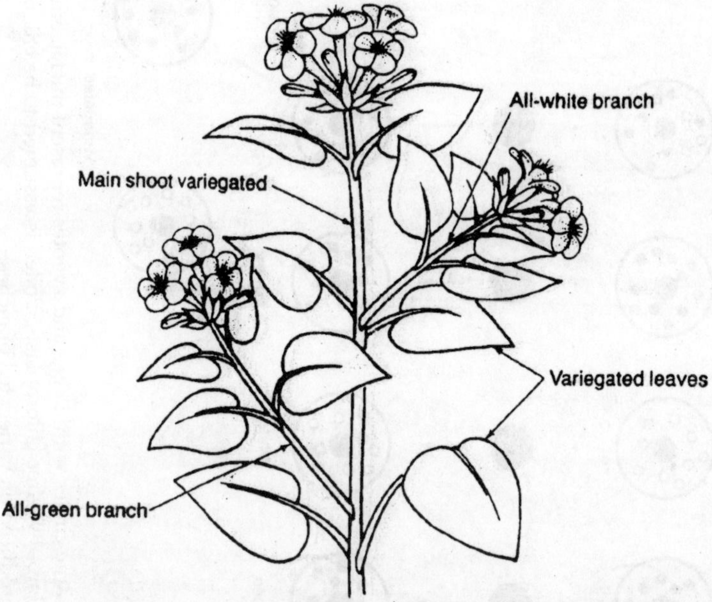

Fig. 1. Leaf variegation in *Mirabilis jalapa*. Shoots that are all green, all white and variegated form on the same plant. Flowers form on all three kinds of shoots.

Table 1
Results of Crosses of Variegated Four O'Clock Plants

Phenotype of branch bearing egg parent	*Phenotype of branch bearing pollen parent*	*Phenotype of progeny*
White	White	White
White	Green	White
White	Variegated	White
Green	White	Green
Green	Green	Green
Green	Variegated	Green
Variegated	White	Variegated, green or white
Variegated	Green	Variegated, green or white
Variegated	Variegated	Variegated, green or white

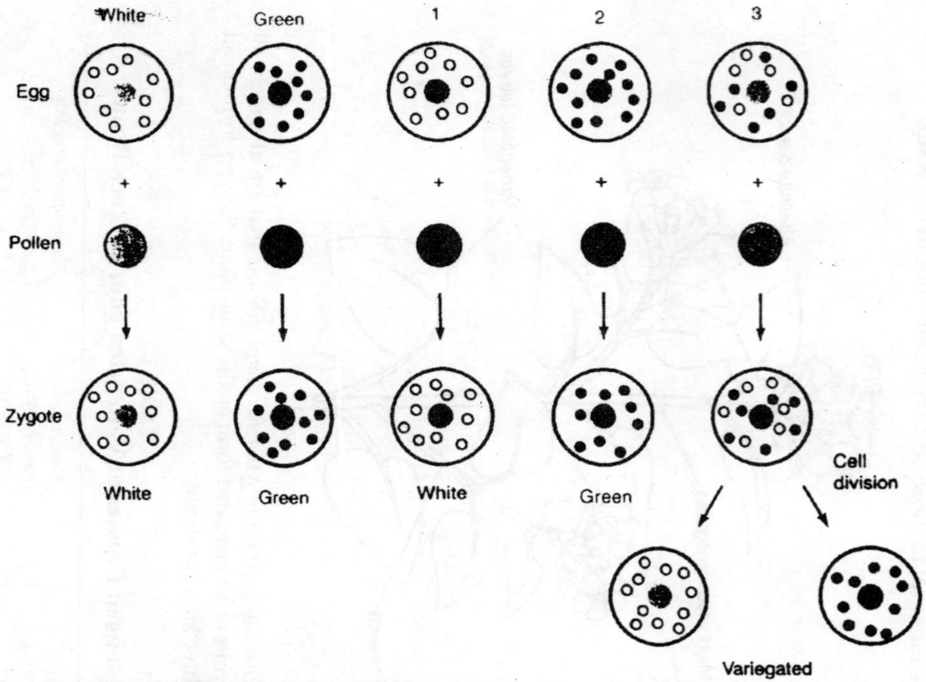

Fig. 2. Hypothesis explaining variegation in *Mirabilis.* Large circles within eggs and zygotes represent nuclei. Small red circles are chloroplasts; small open circles are chlorophyll-free chloroplasts. Pollen is assumed to be chloroplast-free. 1, 2 and 3 represent the three types of eggs produced on variegated branches.

The explanation for the phenomenon is fairly straightforward and is based on three points. (1) Green colour depends on the presence of chloroplasts, and pollen contains no chloroplasts. (2) Segregation of chloroplasts into daughter cells is not determined by any feature of the mitotic apparatus, but by cytoplasmic division. (3) The cells of yellow-white tissue contain chloroplasts that are free of chlorophyll.

These points allow one to draw the model shown in figure. The flowers on green branches produce eggs with a full complement of chloroplasts, so the zygotes have a normal amount of chloroplasts, and all progeny cells will be green. Flowers on white branches produce eggs containing only chlorophyll-free chloroplasts. The eggs formed by flowers on variegated branches are of three types, those containing chlorophyll, those lacking chlorophyll, and those containing both chloroplasts and chlorophyll-free chloroplasts. The third class yields variegated plants, because either during gamete formation or in the development of the plant some type of cytoplasmic segregation occurs that sorts out normal chloroplasts and chlorophyll-free chloroplasts. The molecular mechanism for this sorting is unknown, but there is no reason to believe that it is anything other than random segregation during cell division or gamete formation.

Male Sterility in Plants

Male sterility in plants can be controlled by the nuclear genes or cytoplasm or by both. Therefore, broadly there are atleast three different mechanisms of the control of male sterility in plants. These three types would be briefly discussed in this section.

Genetic male sterility. In this type, sterility is controlled by a single gene and is recessive to fertility, so that the F_1 individuals would be fertile. In the F_2 generation, the fertile and sterile individuals will segregate in 3 : 1 ratio.

Cytoplasmic male sterility. In several crops like maize, cytoplasmic control of male sterlity is known. In such cases if the female parent isd male sterile, the F_1 progeny would always be male sterile, because the cytoplasm is mainly derived from the egg obtained from the male sterile female parent.

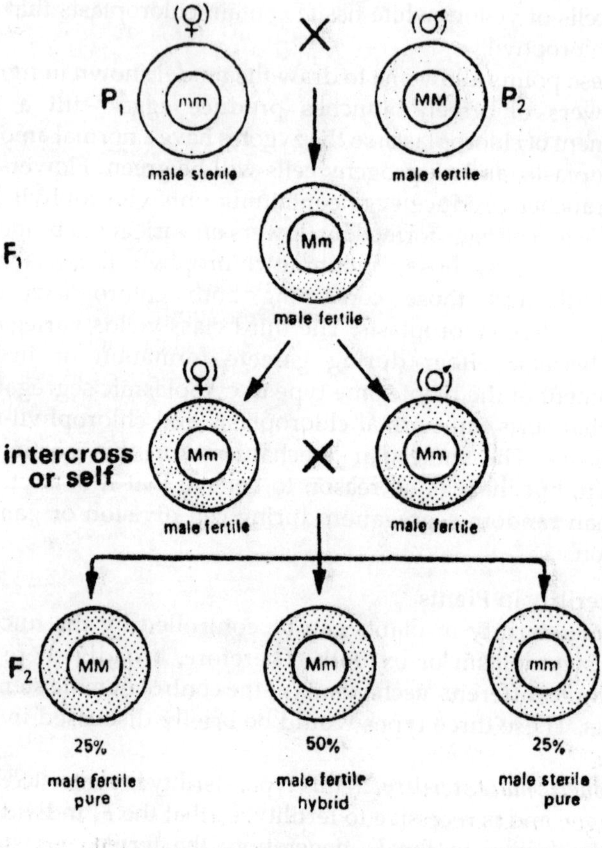

Fig. 3. Inheritance of genetic male sterility.

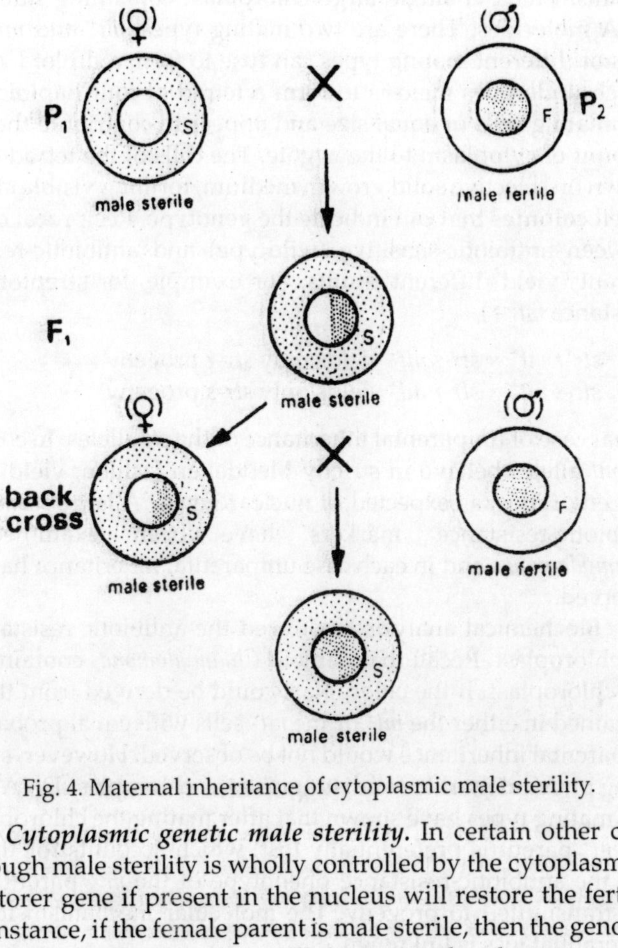

Fig. 4. Maternal inheritance of cytoplasmic male sterility.

Cytoplasmic genetic male sterility. In certain other cases, although male sterility is wholly controlled by the cytoplasm, but a restorer gene if present in the nucleus will restore the fertility. For instance, if the female parent is male sterile, then the genotype (nucleus) of the male parent will determine the phenotype of the F_1 progeny. The male sterile female parent will have the recessive genotype (rr) with respect to restorer gene. If the male parent is RR, the F_1 progeny would be fertile (Rr). On the other hand, if male parent is rr, the progeny would be male sterile. If the f_1 individual (Rr) is testcrossed with rr male, 50 per cent fertile and 50 per cent male sterile progeny would be obtained.

Drug Resistance in Chlamydomonas

One of the most complete studies of cytoplasmic inheritance is with the unicellular alga *Chlamydomonas reinhardii.* Cells of this organism have a single large chloroplast containing numerous DNA molecules. There are two mating types, mt^+ and mt^-, and cells of different mating types can fuse to form a diploid zygote, which undergoes meiosis to form a tetrad of four haploid cells. The mating cells of equal size and appear to contribute the same amount of cytoplasm to the zygote. The cells of the tetrad can be grown on selective solid growth medium, forming visible clusters of cell **colonies** that can indicate the genotype. Reciprocal crosses between antibiotic-sensitive (wildtype) and antibiotic-resistant mutants yield different results. For example, for streptomycin-resistance (*str-r*)

str-r mt^+ × *str-s* mt^- yields only *str-r* progeny
str-s mt^+ × *str-r* mt^- yields only *str-s* progeny

a clear case of uniparental inheritance of the *str* alleles. In contrast, the *mt* alleles behave in strictly Mendelian fashion, yielding 1:1 progeny ratios, as expected of nuclear genes. A large number of antibiotic-resistance markers have been examined in *Chlamydomonas* and in each case uniparental inheritance has been observed.

Biochemical analysis has traced the antibiotic resistance to the chloroplast. Recall that haploid *Chlamydomonas* contains only one chloroplast. If the chloroplast could be derived from the one contained in either the mt^+ or the mt^- cells with equal probability, uniparental inheritance would not be observed. However, studies using physical markers to distinguish the chloroplast DNA of the two mating types have shown that after mating the chloroplast of the mt^- parent is preferentially lost, which accounts for the fact that the antibiotic-resistance phenotype of the mt^+ parent is the one transmitted to progeny. The molecular mechanism for this preferential loss is unknown.

About five per cent of the progeny from *Chlamydomonas* crosses do not exhibit cytoplasmic segregation. These progeny have retained both parental chloroplasts, although they ultimately segregate in later cell divisions. These cells have proved to be quite valuable, because genetic recombination can occur between the DNA in the two chloroplasts. Analysis of the recombination frequencies for the various phenotypes has allowed the construction of fairly detailed maps of the chloroplast genome.

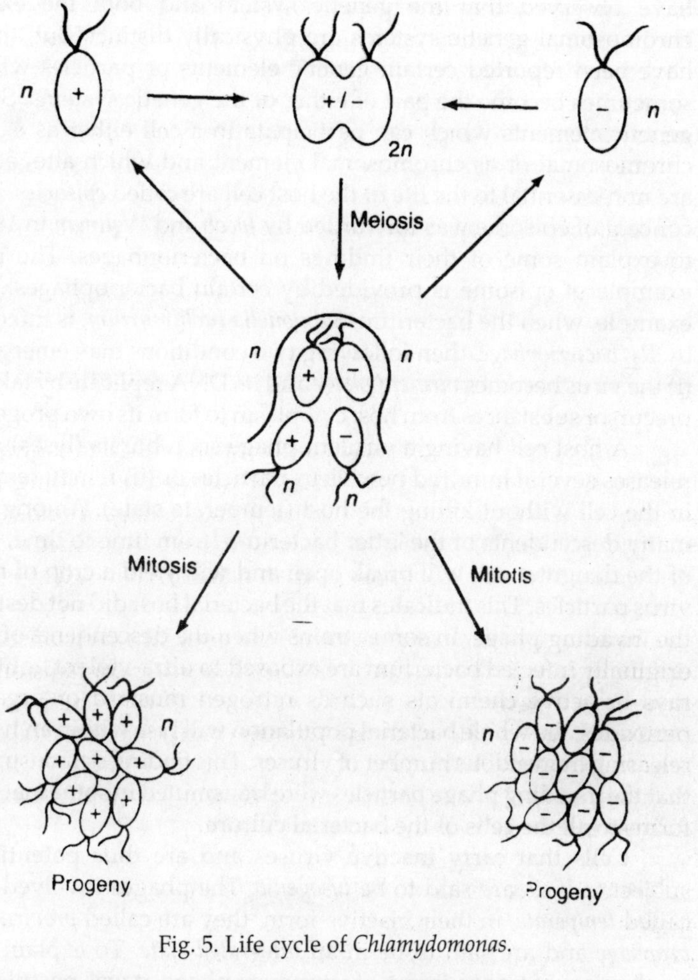

Fig. 5. Life cycle of *Chlamydomonas.*

Non-Chromosomal Inheritance in Prokaryotes

In prokaryotes (i.e., bacteria) many cases of non-chromosomal heritance have been reported. In them, the extra-nuclear inheritance is performed by (i) episomes and (ii) plasmids.

1. Episomes. During the discussion of extra-chromosomal inheritance by normal cell constituents (e.g., chloroplast mitochondria, etc.) and infective symbiotic micro-organism which appear as cellular inclusions (*e.g.*, sigma, kappa, mμ particles), we have observed that the genetic system and both the extra-chromosomal genetic systems are physically distinct. But, there have been reported certain genetic elements or particles which sometimes become the part of either of the genetic systems. Such genetic elements which can participate in a cell either as extra-chromosomal or as chromosomal element and which altogether are non-essential to the life of the host cell are called *episomes.* The concept of episome was forwarded by *Jacob* and *Wollman* in 1958, to explain some of their findings on bacteriophages. The best example of episome is provided by certain bacteriophages. For example, when the bacterium, *Salmonella typhimurium,* is infected by P_{22} *bacteriophage,* then following two conditions may emerge—(i) the virus becomes *virulent phage* and its DNA replicate by taking precursor substances from host cytoplasm to form its own progeny.

A host cell having a virulent phage soon bursts (lyses) and releases several hundred new virus particles or (ii) it may remain in the cell without killing the host (temperate state). Among the many descendents of the latter bacterium, from time to time, one of the daughter cell will break open and will yield a crop of new virus particles. This indicates that the bacterial host did not destroy the invading phage. In some strains when the descendents of the originally infected bacterium are exposed to ultra-violent light, X-rays or active chemicals such as nitrogen mustard or organic peroxides, the whole bacterial population will lyse within an hour, releasing tremendous number of viruses. This finding demonstrates that the invading phage particles were transmitted in some inactive forms to all the cells of the bacterial culture.

Cells that carry inactive viruses and are thus potentially subject to lysis are said to be *lysogenic.* The phages involved are called *temperate,* in their inactive form, they are called *provirus* or *prophage* and are said to be in an *integrated state.* To explain the mechanism of transmission of temperate phage, it was postulated that the DNA or genome of temperate virus becomes attached to the genome of bacterial cell and then behaves as a chromosomal constituent of the host. In the prophage state (inactive state), the genome of temperate phage replicates alongwith the bacterial chromosome and is transmitted to all daughter bacterial cells. In some of the daughter cells, it enters the extra-chromosomal state, by detaching from the host's genome and becomes virulent. In the extra-chromosomal phase, the genetic material of virus has two

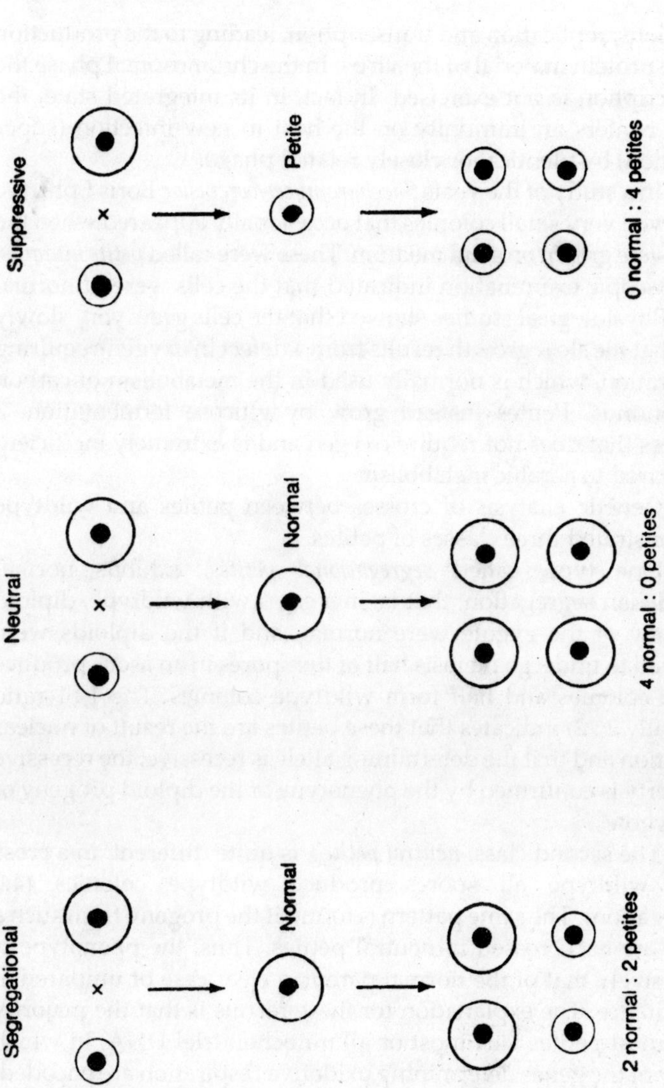

Fig. 6. Behaviour of petite mutations in genetic crosses. A small circle represents a petite cell, a red nucleus contains an allele leading to petite formation, and a shaded cell has normal mitochondria.

functions: replication and transcription, leading to the production of the protein material of the virus. In the chromosomal phase the transcription is not exercised. In fact, in its integrated state, the virus confers an immunity on the host to new infection (super infection) by identical or closely related phage.

In a study of the yeats *Saccharomyces cerevisiae* Boris Ephrussi observed very small colonies that occasionally appeared when the cells were grown on solid medium. These were called *petite mutants.* Microscopic examination indicated that the cells were of normal size. Physiological studies showed that the cells grew very slowly and that the slow growth results from a defect in oxygen-requiring respiration, which is normally used in the metabolism of carbon compounds. Petites instead grow by glucose fermentation, a process that does not require oxygen and is extremely inefficient compared to aerobic metabolism.

Genetic analysis of crosses between petites and wildtype demonstrated three classes of petites.

One type, called *segregational petites,* exhibits normal Mendelian segregation; that is, in a cross with wildtype, diploid progeny of the zygote were normal, and if the diploids were allowed to undergo meiosis half of the spores in an ascus produce petite colonies and half form wildtype colonies. The 1 : 1 ratio (actually 2 : 2) indicates that these petites are the result of nuclear mutation and that the determining allele is recessive; the recessive property is confirmed by the phenotype of the diploid progeny of the zygote.

The second class, *neutral petites,* is quite different: in a cross with wildtype all spores produce wildtype colonies (4:0 segregation). The same pattern is found if the progeny from such a cross are backcrossed to neutral petites. Thus, the phenotype is exclusively that of the normal parent, a clear case of uniparental inheritance. The explanation for these results is that the majority of neutral petites lack most or all mitochondrial DNA, in which many of the genes determining oxidative respiration are encoded. When a neutral petite cell mates with a wildtype cell, the cytoplasm of the latter is the source of the normal mitochondrial DNA in the resulting progeny spores.

The third class, *suppressive petites,* is not well understood. Crosses between these petites and wildtype cells yield petite diploids and mutant ascospores that form petite colonies (0 : 4

segregation). This again shows uniparental inheritance, but in this case the petite condition appears to act as a dominant inhibitor (suppressor) of the activity of the wildtype mitochondria contributed to the cross. Suppressive petites have deletions in their mitochondrial DNA, but not as extensive as the deletions in neutral petites.

Petites arise at a frequency of roughly 10^{-5} per generation. The petite phenotype is the result of large deletions in mitochondrial DNA. For unknown reasons, yeast mitochondria frequently fuse and fragment; this may be the cause of the occasional deletions of DNA. Production of petites is apparently a result of segregation of aberrant mitochondrial DNA from normal DNA and the further sorting out of mutant mitochondria during cell division.

A maternally inherited mutant in *Neurospora*, called *poky*, resembles the yeast petite mutants in being slow-growing and defective in mitochondrial function. In *Neurospora*, unlike yeast, it is possible to make a cross such that one parent (called the female, though there is no true differentiation of the sexes) contributes most of the cytoplasm to the progeny. Maternal inheritance of *poky* is shown by the different results of the following reciprocal crosses:

poky female × wildtype male → All *poky* progeny
poky male × wildtype female → All wildtype progeny

The biochemical deficiency in *poky*, which is controlled by a gene in the mitochondrial DNA, is in the protein-synthesizing system of the mitochondria (mitochondria use a different protein-synthesizing system from the nuclear genes, whose protein products are synthesized in the cytoplasm). The result is the inability of the cell to make proteins that are only synthesized by the mitochondrial system, which leads to the inability to carry out oxidative metabolism. As in the case of yeast *petites*, *poky* mutants are forced to grow via inefficient fermentation pathways.

Cytoplasmic Inheritance in Animals

Several cases of cytoplasmic inheritance in animals have been studied. Some common examples are given under the following sub-heading:

Eye Colour in Flour Moth

The Ephestia. The expression of colour in the recessive genotype in the larvae of *Ephestia* (flour moth) which is often cited as an example of cytoplasmic inheritance is not also a true instance of this kind. What actually happens in this moth is that individuals having *K* either in homozygous or heterozygous condition synthesize kynurenine, a coloured substance in tryptophan metabolism. In the mutant (*k/k*), the substance is not produced and the resultant larvae are colourless. In the backcross, therefore, between heterozygous coloured month (*K/k*) as female and colourless male (*k/k*), half of the individuals should be coloured and half without any colour. Initially, however, all the larvae show colour but later, in half of them colour fades out and the pupae and adults that develop from these larvae do not show any colour.

It is not difficult to understand the appearance of kynurenine in *k/k* larvae, if we assume that any substance produced by the gene remains diffused in the cytoplasm. Therefore, whatever be the genotype of eggs, *K* or *k* resulting from *K/k* heterozygote, the cytoplasm of the egg contains initially kynurenine because the cytoplasm of the heterozygote contains this substance. Later on, this substance used up in *k/k* individuals. Since there is no dominant gene to replenish kynurenine, the larvae which are initially coloured become colourless.

Cytoplamic Inheritance in Ambystoma Meicanum

In axolotl (*Ambystoma meicanum*) a lethal gene O has been discovered which has a maternal effect. The gene O is recessive, so that heterozygote individuals (with genotype $^{+}$O) are completely normal. Homozygous animals (with genotype OO) produced by a cross between two heterozygous parents ($^{+}$O) develop normally in the early stages but in later life show a slight retardation of growth. Their regeneration capacity is severely reduced so that amputated legs are not regenerated as would occur in normal axolotls.

The homozygous (OO) males are sterile—their testes are poorly developed and spermatogenesis does not go beyond the spermatogonial stages. The homozygous (OO) females on the other hand, produce eggs capable of fertilisation and normal cleavage, but at the onset of gastrulation the development is retarded and embryos normally die. This abnormal course of development is in no now way affected by the genotype of the spernatozoon: the spermatozoon may carry the normal +gene or a mutant lethal O gene (if the male is hetezygous $^{+}$O). The arrest in development is in both cases exactly the same: this shows that the particular type of abnormal early development does not depend on the genes present in the cells of the developing embryo ($^{+}$O), but exclusively on the OO genes of the mother (maternal effect), Briggs and Justus (1967) found that eggs of a OO female lack a protein-like substance the "corrective factor" which is necessary for normal development and is synthesized by normal genes + of homozygous or heterozygous normal females (++ or $^{+}$O) during oogenesis.

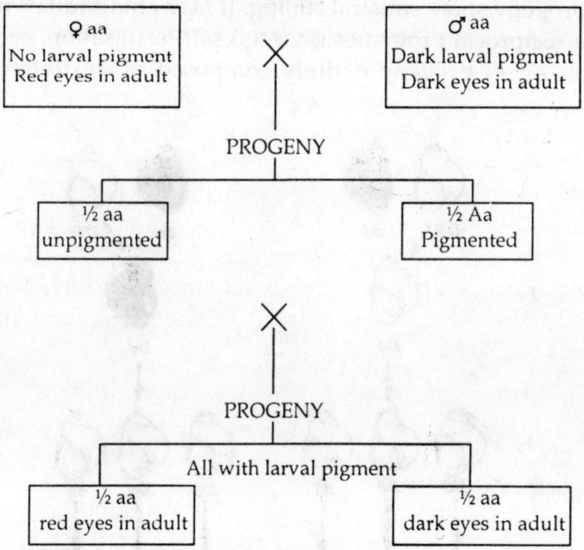

Fig. 7. Maternal effect of the gene A in reciprocal crosses in Ephestia *kühneilla*.

Maternal Effect

The influence of maternal substances on the phenotype of the developing organism is referred to as *maternal effect*. These maternal substances, such as messenger RNAs, are made in the premeiotic oocyte under the direction of maternal genes, and they therefore reflect the maternal genotype and not the zygotic genotype. When the zygote and developing organism are observed, however, zygotic genes seem to be responsible for the expressed phenotype. If the inheritance of such characteristics can be followed over several generations, maternal effect is usually clearly demonstrated and distinguished from true extranuclear inheritance.

In the snail *Limnea peregra,* right-handed (dextral) coiling of the shell is determined by a dominant allele *D,* and left-handed (sinistral) coiling by the recessive allele *d* of a single gene. The animal is hermaphroditic and is capable of self-fertilisation as well as cross-fertilisation. When homozygous dextral (*DD*) are crossed with sinistral (*dd*) males, all the progeny are dextral; but in the reciprocal cross of sinistral *(dd)* females and dextral *(DD)* males, all the progeny show sinistral coiling. If *Dd* F_1 individuals in each of these reciprocal progenies undergo self-fertilisation, each one produces an F_2 progeny entirely composed of dextrally coiled snails.

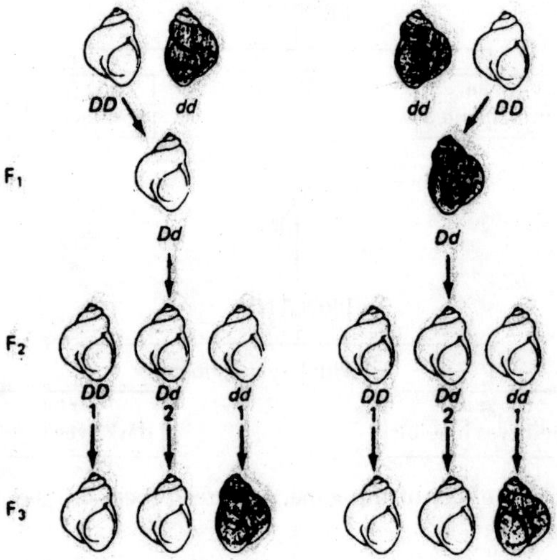

Fig. 8. Maternal effect in the snail *Limnea peregra.* Direction of shell coiling depends on the mother's genotype rather than on the genotype of the individual itself. Dextral, or right-handed, coiling is dominant over sinistral, or left-handed, coiling; but the expected F_1 and F_2 phenotypic ratios are delayed one generation and appear in the F_2 and F_3 generations, respectively.

This seems to be non-Mendelian inheritance, but its true nature is clearly revealed when each F_2 individual undergoes self-fertilisation to produce F_3 progeny. Three-fourths of the F_2 snails produce dextral progeny and one-fourth of the F_2 snails produce an F_3 generation composed entirely of sinistral types. The appearance of the same phenotypic ratio of 3 dextral: 1 sinistral in reciprocal progenies points to single-gene Mendelian progenies points to single-gene Mendelian autosomal inheritance, but this phenotypic ratio appears in the F_3 generation instead of the F_2 generation.

The reason for the delayed expression of Mendelian phenotypic ratios is that substances produced in the egg cytoplasm by the maternal genotype govern the symmetry of the first cleavage division of the fertilized egg, or zygote. Once this division has taken place, the direction of coiling (and of the whole body) is established and lasts the lifetime of the animal. If the maternal genotype is *DD*, all zygotes will be dextral: if the maternal genotype is *dd*, all zygotes will be sinistral, regardless of the male parental genotype. Each F_1 *Dd* female in reciprocal progenies, whether dextral or sinistral in phenotype, has the dominant *D* allele. Every egg produced by these *Dd* females has the *D* gene product, which directs dextral coiling in F_2 zygotes developed when these eggs are fertilized. When the F_2 snails self-fertilize, those with the *dd* genotype give rise to sinistral progeny whereas females with *DD* or *Dd* genotypes will produce dextral F_3 progeny.

The overall result shows classical Mendelian inheritance, delayed one generation. If the pattern had been truly extranuclear, differ ences would have been evident in reciprocal progenies in every generation and not just in one or two generations of a lineage. In extranuclear inheritance the progeny continue to resemble only one type of the two parents and the extranuclear trait is transmitted indefinitely in subsequent generations. The inheritance of phenotypic traits can therefore be judged by various criteria using genetic methods and molecular analyses. If the pattern of inheritance is extranuclear, the responsible extranuclear genome can be identified in many cases as belonging to organelles, such as mitochondria and chloroplasts, or to viruses or plasmids. Investigation of some new extranuclear inherited trait in an organism is therefore guided by available methods through which the extranuclear factor can be located. It can then be explored in greater detail in that host system.

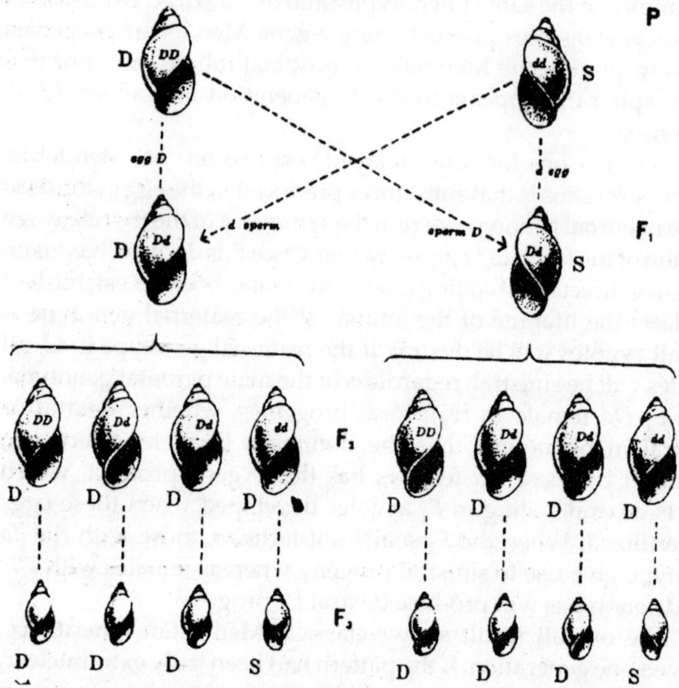

Fig. 9. The normal inheritance of dextral and sinistral coiling in *Limnaea peregra*. D, dextral phenotype; S, sinistral phenotype; *D*, gene for dextrality; *d*, gene for sinistrality. Note: This species is hemaphroditic and may reproduce either by crossing or by self-fertilisation.

Virus-like Inclusions and Infective Particles

The study of cytoplasmically transmitted, self-reproducing substances has attracted much attention in recent years. Three cases of special interest are the milk factor, a maternally transmitted agent leading to the formation of mammary tumors in certain strains of house mice; the kappa particles in Paramecium; and the sigma factor for carbondioxide sensitivity in Drosophila.

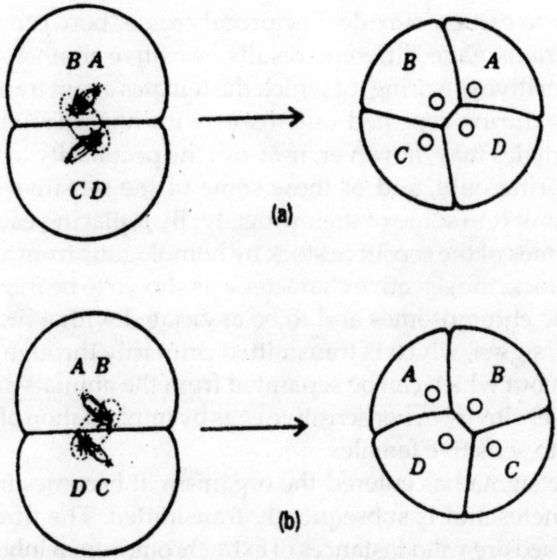

Fig. 10. Coiling pattern in a snail as determined by the orientation of the spindles for the second cleavage division. Cleavage leading to sinistral pattern is shown by *a* and that leading to dextral pattern by *b*.

The Milk Factor

In outcrossing mice from a long-inbred strain in which nearly all females developed a particular type of mammary cancer, Bittner found that cancer susceptibility was maternally transmitted. Moreover, by using females from the high-cancer strain as foster mothers for new-born mice from noncancer strains and normal mothers as nurses for young from the high-cancer strain, it was shown that the substance that induces cancer-susceptibility is transmitted through the mother's milk, whence it has become known as the "milk factor". Recent evidence suggests that it is probably a virus, which, since it is found also in the seminal vesicles, may be transmitted like an infection by males of certain high-tumor strains. Whether it can reproduce itself indefinitely apart from a certain genetic constitution of the host is not yet known. It may be lost spontaneously from strains carrying it and thus is not necessarily a continuously transmitted entity.

The Sigma substance in Drosophila

In *Drosophila melanogaster*, L'Heritier and Teissier found a true-breeding stock that differed sharply from others in its high sensitivity to carbon dioxide. Reciprocal crosses between this and normal strains gave different results, sensitive mothers having mostly sensitive offspring, of which the females again transmitted sensitivity during repeated outcrosses with nonsensitive males. Sensitive males may, however, transmit the peculiarity to a few of their offspring only, and of these some of the sensitive females may transmit it to some of their progeny. By replacing each of the chromosomes of the sensitive stock by homologous from a normal resistant stock, the sensitive character was shown to be transmitted outside the chromosomes and to be associated with a heat-labile substance, sigma, which is transmitted primarily through the egg cytoplasm but which can be separated from the animals and used to induce sensitivity in nonsensitive eggs by implantation of normal ovaries into sensitive females.

Once sigma has entered the organism, it becomes included in the gametes and is subsequently transmitted. The three cases just described are valid instances of extrachromosomal inheritance. It is to be noted that none of the cytoplasmic particles involved in these cases had been known as components of the cytoplasm; they were discovered in the course of investigating the extrachromosomal transmission of phenotypic traits.

Merogonic Hybrids

The relative roles of cytoplasm and nuclear heredity have also been studied by methods quite different from those hitherto considered. These involve the production of merogonic hybrids, having egg cytoplasm from one parent species and a sperm nucleus from another. The classic experiments in this field were performed by Theodor Boveri, who removed the nucleus from the eggs of a sea urchin of one species and artificially fertilised them by sperm of a species with markedly different characters. Although development did not persist beyond the pluteus stage, the merogonic hybrids seemed to resemble the species that furnished the sperm, indicating that the nucleus controlled development even when surrounded by cytoplasm from a very different source. Boveri's results were open to serious question, however, since there is evidence that the nuclear material was not entirely removed. In repeating these experiments, Hörstadius did indeed find differences between the reciprocal merogonic hybrids, indicating predetermination of the egg cytoplasm by its own genes.

The work of Hadorn, however, indicates that for certain traits the male nucleus may be without effect in some merogonic tissues. Two species of Triton (*T. palmatus* and *T. Cristatus*) differ clearly in the character of the epidermis. Hadorn fertilized the egg of *palmatus* with the sperm of *cristatus* and succeeded in removing the *palmatus* nucleus before nuclear fusion. The embryo (thus haploid) develops only to the blastula stage; but a portion of the presumptive epidermis, if grafted to a normal larva of another species, *T. alpestris*, maintains its identity and develops to the adult state. Here it resembles the epidermis of the parent *palmatuus*, which contributed the cytoplasm, rather than the parent *cristatus*, which contributed the nucleus. For this trait, at least, the cytoplasm rather than the nucleus seems to have the decisive effect, a result quite the opposite from Boveri's.

The Spread of Pigment in the Skin

The problem of cytoplasmic transmission has been attacked in still a different way. For many years, it has been known that when pieces of skin are transplanted from black to white areas of black-and-white spotted guinea pigs, the graft continues to produce black pigment in the skin (but not in the hairs) and, with time, black pigment spreads to the formerly colourless areas surrounding the graft. Billingham and Medawar showed that when areas altered in this way are transplanted to white areas, pigment production and spread to new white areas continue through several serial propagations. The effect resembles the spread and increase of an infective principle which passes from cell to cell after the nuclei of the original black cells are no longer present.

Billingham and Medawar found in the white areas of guinea pig skin dendritic cells like the melanocyte cells of the coloured areas in which the pigment is formed. They assumed that cytoplasmic connections between the black and the white cells formed a bridge over which passed a principle capable of transforming colourless into coloured cells and considered this property as an additional aspect of cellular heredity, important in development and differentiation. The hypothesis apparently does not explain the appearance of black hairs in grafts of white skin into black areas on rats and mice. This occurs by migration of melanoblasts. It is not yet apparent that an additional system of self-propagating elements has to be invoked to explain the spread of skin pigment in cases like those above.

Cytoplasmic Transmission of Symbionts

In eukaryotes variety of cytoplasmically transmitted traits result from the presence of bacteria and viruses living in the cytoplasm of certain cells. One of the best understood is the killer phenomenon found in certain strains of the protozoan *Paramecium aurelia.* Killer strains of *Paramecium* release to the surrounding medium a substance called paramecin, which is lethal to many other strains of the protozoan. The killer phenotype requires the presence of particles, called *kappa,* whose maintenance is dependent on a dominant nuclear gene *K. Paramecium* is a diploid protozoan that undergoes sexual exchange via conjugation. The conjugation cycle is an unusual one. Initially the cell has two diploid micronuclei (many protozoans contain micronuclei—small nuclei—and a macronucleus—a large nucleus with a specialised function—but only the micronuclei are relevant to the genetic processes we are describing.

Two cells come into contact, and prior to any conjugation the two micronuclei in each cell undergo meiosis, forming eight micronuclei in each cell. Seven of the micronuclei disintegrate, leaving each cell with one micronucleus, which then undergoes one mitotic division. The cell membrane between the two cells breaks down slightly and the cells exchange a single micronucleus, after which the nuclei fuse to form a diploid nucleus. After nuclear exchange the two new cells (the exconjugants) are genotypically identical. In this sequence of events only the micronuclei are exchanged, without any mixing of cytoplasm. Single cells of *Paramecia* also occasionally undergo an unusual nuclear phenomenon called *autogamy.* Meiosis occurs and again all of the micronuclei disintegrate except one. This surviving nucleus then undergoes mitosis and nuclear fusion to recreate the diploid state. The critical point about autogamy is that if the cell was initially heterozygous, the newly formed diploid nucleus is homozygous because it was derived from a single haploid meiotic product. In a population of cells undergoing autogamy for surviving micronucleus is selected randomly, so for any pair of alleles half of the new cells will contain one allele and half will contain the other. Under certain conditions conjugation, which normally involves *only* an exchange of micronuclei at the surface of contact between the two cells, also includes cytoplasmic mixing followed by autogamy.

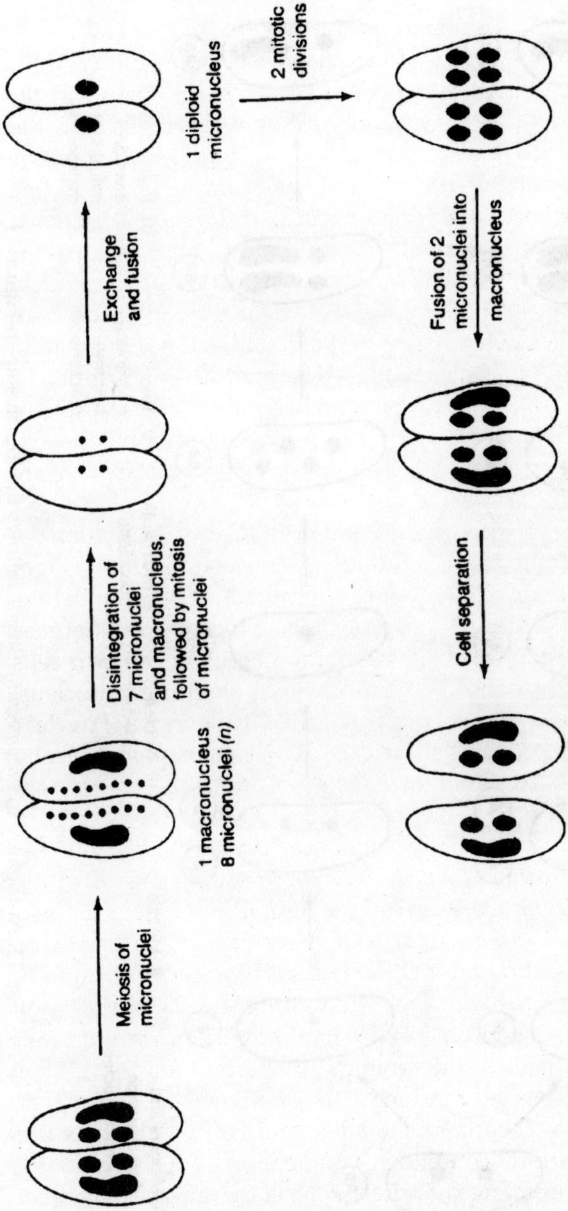

Fig. 11. Conjugation in *Paramecium.*

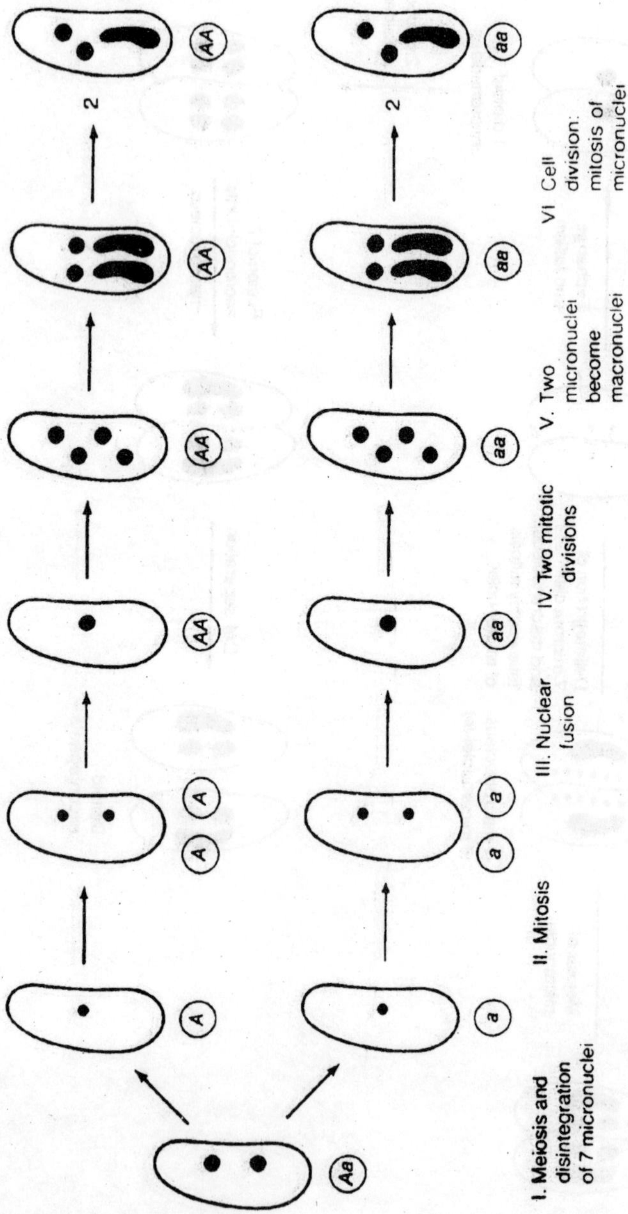

Fig. 12. Autogamy in *Paramecium*. Step 1 follows the fourth step in Figure 11.

A comparison of the phenotypes following conjugation of killer (*KK*) cells and sensitive (*kk*) cells with and without cytoplasmic mixing and autogamy yields the evidence for cytoplasmic inheritance of the killer phenotype. Without cytoplasmic mixing the expected 1 : 1 ratio of killer to sensitive cells results. When autogamy occurs, the production of homozygosity converts the *Kk* killer cells to one *KK* killer cell and one *kk* sensitive, which does not express the killer phenotype even if kappa is present because there is no *K* allele. The situation is quite different when cytoplasmic mixing occurs. In this case, conjugation yields two killer cells since each cell has kappa particles derived from the cytoplasm of the killer parent. The results of autogamy in each of the exconjugants, namely, the production of one killer cell and one sensitive cell, indicates that the exconjugants were *Kk* heterozygotes.

Microbiological studies of killer strains have shown that kappa particles are actually a symbiotic bacterium called *Caedobacter taeniospiralis.* These bacteria produce paramecin. Why the killer strains are not killed by paramecin has not yet been determined. Another example of a symbiont-related cytoplasmic effect is a condition in *Drosophila* called *maternal sex ratio,* which is characterised by the production of few male progeny. The daughters of sex-ratio females pass on the trait, whereas the occasional sons that are produced do not. Cytoplasm taken from the eggs of sex-ratio females transmits the condition when injected into females from unaffected cultures of the same or other *Drosophila* species. A bacterium has been isolated from the cytoplasm of sex-ratio females; when this bacterium is allowed to infect other *Drosophila* females, they acquire the sex-ratio trait. The causative agent of the sex-ratio condition is not the bacterium itself, but a small virus that multiplies in the bacterium. This virus, when released by the bacterial cells, kills most male *Drosophila* embryos. Why female embryos are not killed by the virus is unknown.

Milk Factor in Mice

Bittner discovered a case of cytoplasmic inheritance in mice. He had a strain of mice in which almost all developed breast cancer. From these he took some females and bred them to males which did not inherit this tendency. 90 per cent of the offspring of such crosses formed breast cancer. On the other hand, when he took

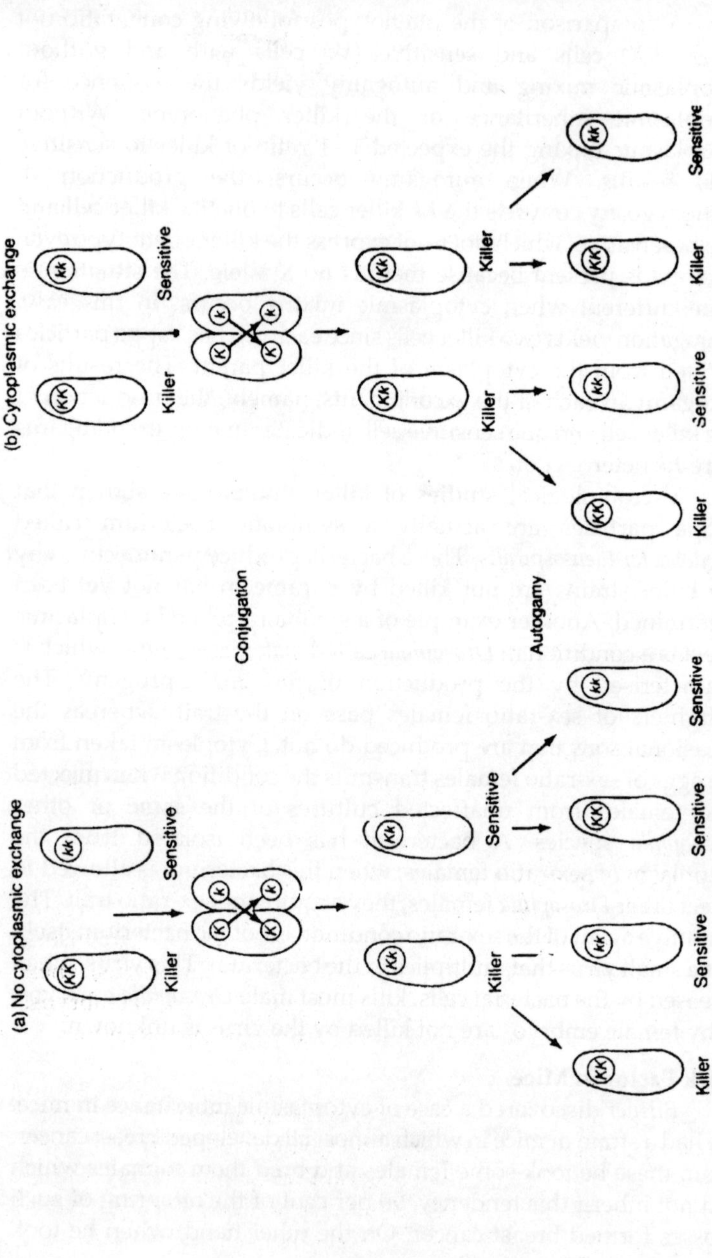

Fig. 13. Crosses between killer (*KK*) and sensitive (*kk*) *Paramecium* with and without cytoplasmic exchange during conjugation. Cells containing kappa are shown in pink. The genotypes of each cell are indicated.

males from the cancer-susceptible strain and crossed them to normal females, *Bittner* found no cases of breast cancer among the offspring. These results pointed out to cytoplasmic inheritance. Late studies pointed out that it was not the case. When he took the new-born off-spring from a cancer strain female and put them with a foster mother from a normal strain for nursing, there were no breast cancer in the offspring. This pointed out that the milk was responsible for transmission of breast cancer. A virus-like *'milk-factor'* was identified and it was discovered that hormones and chromosomal genes play a part by making the mice susceptible to this milk factor.

9

Chromosomal Aberrations

When a chromosome breaks, the two broken ends usually undergo *restitution* or healing, and everything is as before. But this does not always happen, and the result can be the formation of (1) deficiencies, (2) duplications, (3) inversions, or (4) translocations. The first three of these ordinarily involve a single chromosome, while the fourth involves two or more chromosomes.

The results of breakage and rejoining give rise to four different kinds of chromosomal aberrations. There is very strong evidence that the DNA in chromosomes extends as a double helical thread throughout the entire length of the structure. Furthermore, we know that the genes are linearly arranged in the DNA and that at least in some instances groups of genes under the control of common regulatory units are grouped together. It is of great interest then to ask about the genetic consequences of breaks in specific regions of chromosomes. One interesting observation made by George Lefevre on *Drosophila* chromosomes is that perhaps as many as 70 per cent of chromosome break points that have successfully reunited with new neighbouring genes have no detectable genetic effect at all. In other words, only about 30 per cent of healed break points have mutant effects associated with them. This could mean that only a portion of the DNA of chromosomes is informational, and breaks within the regions without information are inconsequential. It could also mean that these regions may be more susceptible to breakage and rejoining than others.

C.S. Lee has suggested that those break points that manage to reunite are likely to be in regions of the DNA that are represented repeatedly in the genome. The reasoning behind this suggestion has to do with the healing process itself rather than with a differential fragility of chromosome regions. The union of broken ends may involve some degree of base pairing at the joint in order for the DNA strands to be properly positioned and stabilised for ligase to be effective in forming covalent bonds between the two chains. It is of course possible that unions can be formed without a perfect match between the bases of the two strands. Experiments with λ phage by Wu and Taylor have shown that the ends of the λ chromosome in its linear form have complementary sequences of nucleotides at the protruding cohesive ends. The λ chromosome goes from the linear to the circular form by the formation of hydrogen bonds between the complementary bases as shown in the figure. The same sort of protruding single strand ends with complementary or mostly complementary sequences may cause the joining of broken ends of eukaryotic chromosomes.

Table 1
Chromosomal Aberrations, the dot Represents the Centromere

Types	*Description*	*Chromosome structure changes*
Normal	Loss of chromosomal segment	1 2 3 4 5 6
Deficiency	Addition of chromosomal segment	1 2 3 4 5
Duplication	Rearrangement of chromosomal	1 2 3 4 5 6 6
Inversion	segments	1 2 5 4 3 6
Translocation	Chromosomal segment exchanged between nonhomologous chromosomes	1 2 3 4 5 6 10 11 12

Broken chromosome ends have been described for many years by cytogeneticists as being sticky. This stickiness may be nothing more than the complementarity that exists between repeated DNA sequences. It is known that in some tissues of many organisms, such as the endosperm of maize kernels, where the broken ends of chromosomes remain unhealed at the time of chromosome replication, that there is often a fusion of the broken ends of the sister chromatids. This results in a dicentric chromosome that will produce a chromosome bridge at the anaphase of the next mitotic division. In plant cells, such a bridge will often break as the cell wall is laid down between newly formed daughter cells. This produces another broken end in each of the nuclei, and the cycle may begin again. This breakage fusion-bridge cycle may be carried on for several cell generations, and it can be used to great advantage for manipulating gene dosage in such cells. As the bridge breaks, it may not do so at the original fusion point and the result will be a duplication of some genes in one cell and their deletion in the other. McClintock used such a scheme to study chromosome organisation and gene regulation in the endosperm of maize.

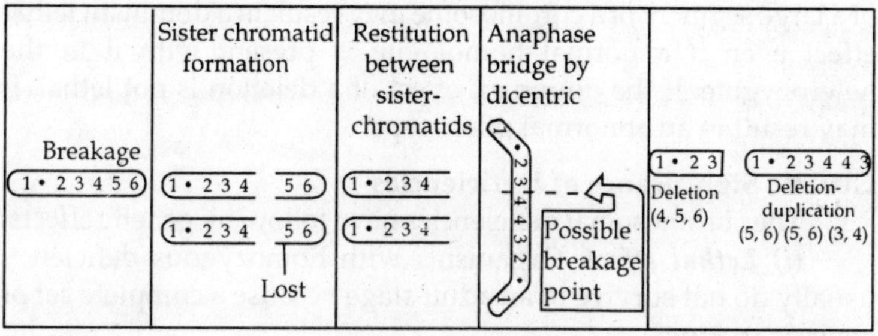

Fig. 1.

Deletions

Deletions are also referred to as *deficiencies*. The terms deficiency and deletion as synonyms, because in either case an acentric fragment is lost. A spindle fiber cannot attach to an acentric fragment during mitosis or meiosis and guide it to one of the poles.

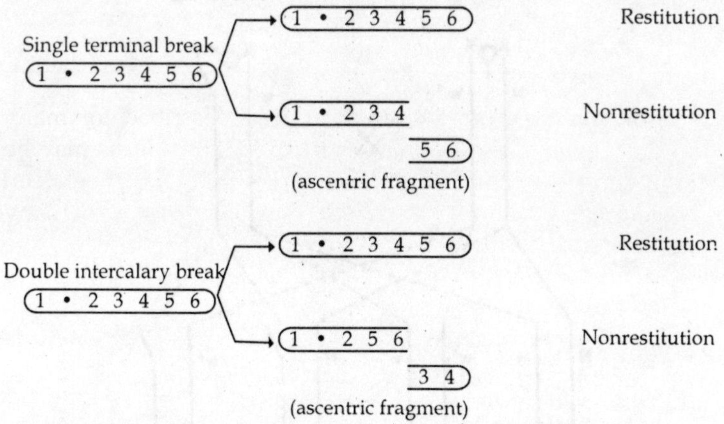

Fig. 2. Two ways in which deletions or deficiencies are formed.

If the DNA in the chromatin that is lost is crucial for viability, sterile gametes or nonfunctional somatic cells may result. Some deficiencies are known that are viable even when homozygous. For example, in *D. melanogaster*, a deficiency that removes the *white* locus produces, when homozygous, fully developed flies that appear normal except for their white eye colour. Most deficiencies, however, are lethal when homozygous, and these may involve no more loss of chromatin than in the example of white locus. A loss of a large segment of a chromosome may result in a dominant lethal effect even if a normal homologue is present with it in the heterozygote. If the dominant effect of a deletion is not lethal, it may result in an abnormal phenotype.

Genetic Significance of Deficiencies

The chromosomal deficiencies have following genetic effects:

(i) Lethal effect. Organisms with homozygous deficiency usually do not survive to an adult stage because a complete set of genes is lacking.

Example. A mutant sex-linked condition called "notch" is lethal in *Drosophila* when hemizygous in males or when homozygous in females. This "notch" phenotype is a sex-linked deletion (deficiency) which acts like a dominant mutation: a deletion at another sex-linked locus behaves as recessive mutation, producing yellow body colour when homozygous.

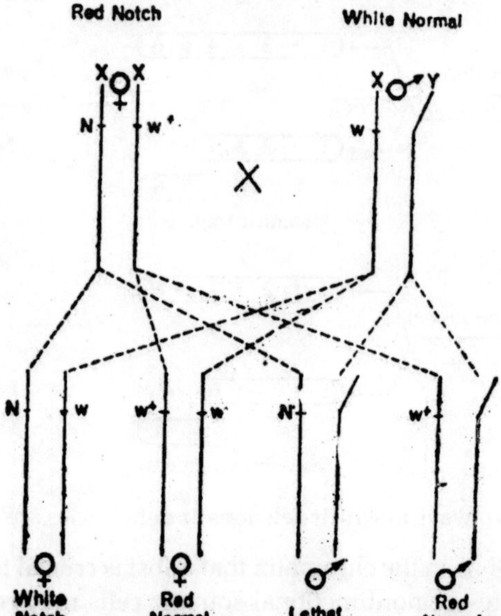

Fig. 3. Segregation involving the deficiency notch in the X-chromosome of *Drosophila melanogaster*. It contains the gene locus for white eyes w (red eyes w_+). The notch type is dominant, but homozygous (in the female) and hemizygous (in male) it is lethal. The w gene in heterozygous notch female is hemizygous (pseudo-dominant).

(ii) Pseudodominance. When an organism heterozygous for a pair of alleles, A and a, loses a small portion of the chromosome bearing the dominant allele (A), the recessive allele (a) on the other chromosome being hemizygous will become expressed phenotypically. The phenotypic expression of recessive allele (a) due to deficiency of dominant allele (A) called **pseudodominance**.

With a deficiency exchange (crossing over) becomes impossible. Disturbance of pairing causes a reduction of exchange in a segment around the deficiency.

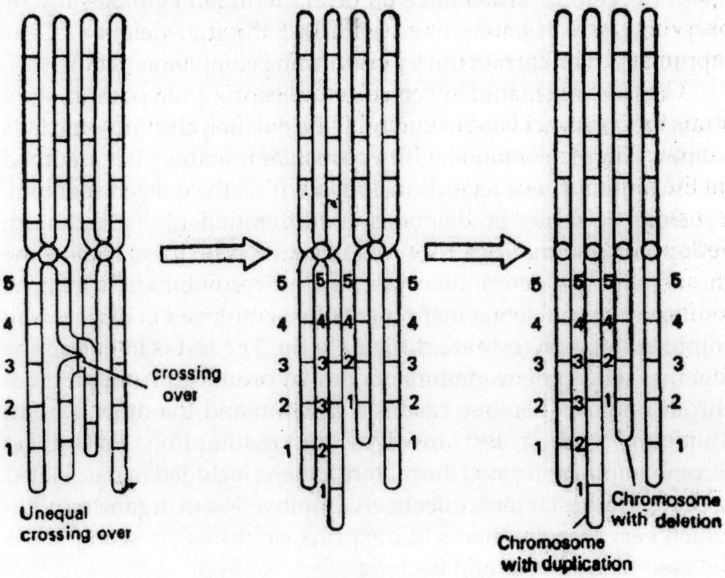

Fig. 4. Duplication and deletion.

Mapping with Deficiencies

Deficiencies are widely used in genetic research to make chromosome maps, particularly for short regions of the chromosomes. In fact, this may be much more efficient than using the crossover values obtained from crosses between heterozygotes if a large number of deletions is available. When two sites are very close together on the chromosome, crossovers between them may be extremely difficult to detect because of the very large number of offspring that have to be classified. This is no great problem with prokaryotes, but it is with eukaryotes like *Drosophila* which would have to be raised by the millions to provide sufficient data to measure recombination values.

An example of the use of deficiencies may be taken from the genetic analysis of the region in *D. melanogaster* between *white* and *zeste*. This region is only 0.75 map units long. Obviously large numbers of flies would have to be counted to determine the order of the sites that are described as being in this region. Figure shows the extent of a series of deletions involving the *white-zeste* and adjacent regions. These have all been identified cytologically by observing which bands have been lost through deletion. Gene mapping is then carried out by performing complementation tests.

Crosses are made between deletion-bearing and point-mutant strains to construct heterozygotes. If the deletion chromosome fails to show complementation with a particular mutation, it is assumed that the mutant gene locus is included within the deleted segment. By using a series of deletions of different lengths and with overlapping boundaries, the order of gene loci on the chromosome can actually be determined just as in a recombination map. A modification of deletion mapping is often employed in cases where complementation tests are difficult to do. The test is to determine whether wild type recombinants can be produced from parental chromosomes when one carries a deletion and the other a point mutation. Such a test involves the assumption that if no recombinants are formed the mutant locus is included in the deleted regions. This test is most effectively employed with organisms with which very large numbers of offspring can be easily scored, as is the case with viruses and bacteria.

Deletions in Human Beings

Deletions are expected to occur in all organisms including humans. Some do not cause early spontaneous abortions, but unfortunately allow the fetus to came to full term and to be born as an abnormal child. One example of this is the *cri-du-chat* or "cat cry" syndrome, which derives its name from the fact that the afflicted infant cries like a cat. Approximately 85 per cent of children with *cri-du-chat* show a heterozygous deletion for the short arm of chromosome 5. The remaining 15 per cent exhibit a translocation of part of chromosome 5, as described in the section on translocations later in this chapter.

Children with *cri-du-chat* syndrome are mentally retarded and atypical in appearance. Generally parents of *cri-du-chat* children are normal, which means that the children arose from zygotes in which one of the participating gametes contained a deleted chromosome. This must have occurred spontaneously in one of the parental germ cells.

Deletions in humans have been useful in locating genes to a particular chromosome. Although the chromosomes of most animals (including humans) do not show clear chromosome banding, special fluorescent staining procedures reveal banding patterns that are useful in identifying chromosomes and relating genes to specific regions of chromosomes.

Karyotypic studies in humans, utilising chromosome banding techniques with various stains and correlating the presence or absence of particular bonds, has permitted the localisation of the MN blood group locus near the centromere of the long arm of chromosome 2. This feat was possible because of the deletion of a small segment of DNA from chromosome 2 of a boy previously reported to be hemizygous (L^M/–) for this locus. The boy was typed as blood type M, and the father as N. The mother was MN, and two MN children and one M child were present in this family. Blood typing studies indicated that the father was L^N/L^N and the child L^M/–, that is, the father had two functional alleles and the child had only one allele at this *L* locus.

Karyotypic analyses of the chromosomes from the child showed a reciprocal translocation between the long arm of chromosome 2 and chromosome 4. This rearrangement was assumed to have occurred *de novo* since the karyotypes of both parents were normal. Measurements of the translocated chromosome 2 of the child suggested that a small chromosomal segment from chromosome 2 was deleted during the rearrangement process. Subsequently, banding patterns have supported the deletion hypothesis based on chromosome measurement in that a segment of DNA from the long arm of chromosome 2 was missing. Therefore, since the serological studies suggested that the boy was hemizygous for the blood group M and a deletion was present, the MN locus was proposed to reside on the long arm of chromosome 2. Deletion mapping of human chromosomes by means of phenotype-karyotype correlations has great potential; however, this is the first locus to be assigned to a well-defined autosomal region by this approach.

Duplications

Duplications are repeats of chromatin segments within a chromosome. Chromosomes containing a duplication therefore possess more than the normal amount of DNA. There are several possible mechanisms for generating duplicate DNA segments in chromosomes. Unequal crossing over between homologous chromatids may yield one chromosome with a duplication and its homologue with a corresponding deletion. The chromatid that bears the duplication will have what is referred to as a tandem repeat. If unequal crossing over occurs the resulting duplicate segments of the tandem repeat will be identical only if the two homologous chromosomes carried identical alleles in the segments.

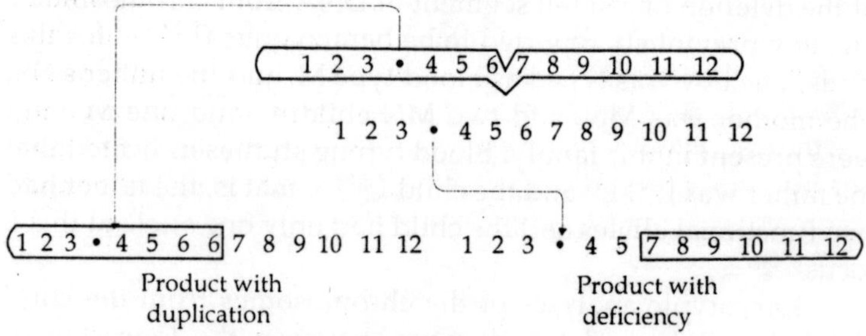

Fig. 5. The results of unequal crossing over.

Duplications and deletions may also be generated by the breakage-fusion-bridge cycle. A chromosome with the normal sequence of 1.23456 has a break between 4 and 5. Since the deleted 5.6 is an acentric fragment it will probably be lost. The replication of the 1.234 fragment, followed by joining of the two broken ends, results in the dicentric chromosome. 1 234–432.1. Since homologous centromeres generally move in opposite directions, an anaphase bridge is formed and breakage will occur. The genetic constitution of the resulting chromosomes depends on where the break occurs.

Table 2
Examples of Repetitive DNA Sequences

Group	*No. repetitive DNA sequences*	*Degree of DNA or protein homology, percent*
Satellite DNA	103–106	80–100
18S–28S Ribosomal RNA	100–600	97–100
5S Ribosomal RNA	100–200	97–100
tRNA	6–400	
Histones	10–1200	87–99
Haemoglobins	≃ 10	< 75–100

The breakage-fusion-bridge cycle that has been found in maize results in a wide variety of duplications and deletions. Note that duplications created in this manner often are displaced, that is, duplicated segments are not positioned side by side on the chromosome.

Duplications play an important role in the regulation of gene activity because the multiple copies of a gene allow large quantities of RNA to be generated quickly. Examples of this type of reiteration are the cistrons coding for the ribosomal RNA's and those coding for histones. Table gives some examples of proteins and RNA's thought to be coded by repetitive DNA sequences.

Genotype	*No. of facets/eye*	*Description*
+/+	779	Homozygous wild-type female
B/+	358	Heterozygous Bar female
B/Y	90	Bar male
B/B	68	Homozygous Bar female
BB/Y	70	Double bar male
BB/+	45	Double Bar female
BB/BB	25	Homozygous double Bar female

Satellite DNA's contain large numbers of repeating sequences which apparently are never transcribed. The nature of the mechanism(s) that generate large numbers of DNA segments in tandem linkage is unknown but it may well be by unequal crossing over. Also unknown are the factors that protect the repeated sequences from mutational changes.

Genetic Significance of Duplications

The duplications have following genetical importance:

1. The duplications of chromosomes are not deleterious to the organism like the deficiency, but, they usually protect the organism from the effect of a deleterious recessive gene or from an otherwise lethal deletion.

2. Large duplications can reduce the fertility as a result of meiotic complication, and in this way reduce their own probability of survival.

3. Relocation of chromosomal material without altering its quantity may result in an altered phenotype, this is called *position effect*.

4. Some duplications are useful in the evolution of new genetic material. In an organism with duplications, because the old genes can continue to provide for the present requirements of the organism, the superfluous genes may be free to mutate to new forms without a loss in immediate adaptability.

Example: The best example of position effect of duplication has been reported in *Drosophila*. The wild type *Drosophila* contains fully developed, oval-shaped compound eyes, while, mutant strains have extrmely reduced, narrow, oblong, *bar-shaped eyes*. Bridges (1936) showed bar eye trait to be associated with the duplication of a segment of X chromosome called section 16A, as observed in salivary gland chromosomes. Each added section 16A intensifies the bar phenotype. However, the narrowing effect is greater if the duplicated segments are on the same chromosome. In bar eye, each added segment narrows the eye still farther, and this effect is enhanced as more duplications occur in one chromosome.

Duplication of Complete Gene System

In considering gene duplications, it is important to remember that a structural gene cannot be considered an entity in itself. The expression of any structural gene is controlled or regulated by DNA not directly involved in coding for proteins. Therefore, the duplication of a structural gene without the corresponding duplication of its regulator may lead to a large array of new phenotypes including lethality.

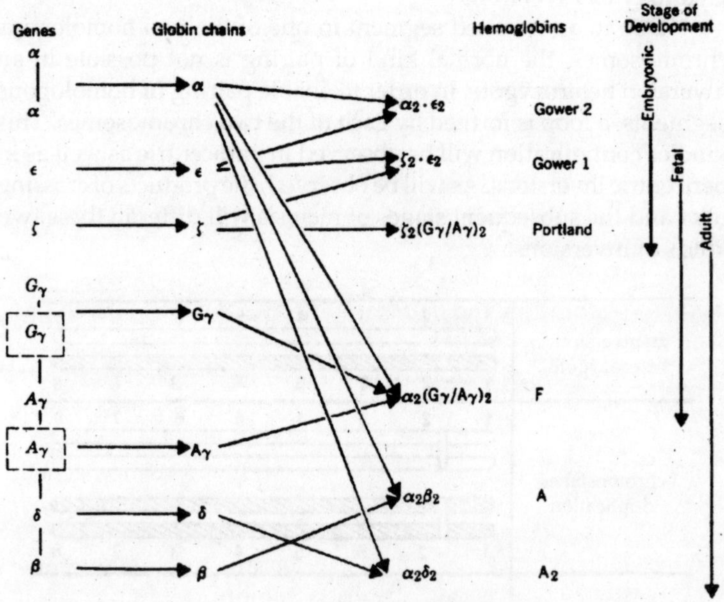

Fig. 6. Genetic control of human haemoglobin. The broken-line boxes indicate the doubt about the number of γ-chain genes per haploid genome.

Inversions

An inversion is produced when there are two breaks in a chromosome and the intercalary segment reunites in the reverse order i.e., the segment rotates at 180°. Let us imagine that chromosome 1 2 3 4 5 6 7 8 gives rise to another chromosome having the order 1 2 7 6 5 4 3 8. The segment 3-4-5-6-7 has rotated here at 180° giving an inverted order of genes 7 6 5 4 3.

The inversions can be of two types—

(1) paracentric inversion and (2) pericentric inversion.

Paracentric inversions are those inversions, where the inverted segment does not include centromere. On the other hand, in a pericentric inversion, the inverted segment includes the centromere. In order to remember these terms and their meaning, one should bear in mind that pericentric means surrounding the centromere or on the periphery of centromere.

Cytology of Inversions

Due to an inverted segment in one of the two homologous chromosomes, the normal kind of pairing is not possible in an inversion heterozygote. In order to enable pairing of homologous segments, a loop is formed by each of the two chromosomes. This kind of configuration will be observed in paracentric as well as in pericentric inversions. As will be observed, the products of crossing over and the subsequent stages of meiosis will differ in these two kinds of inversions.

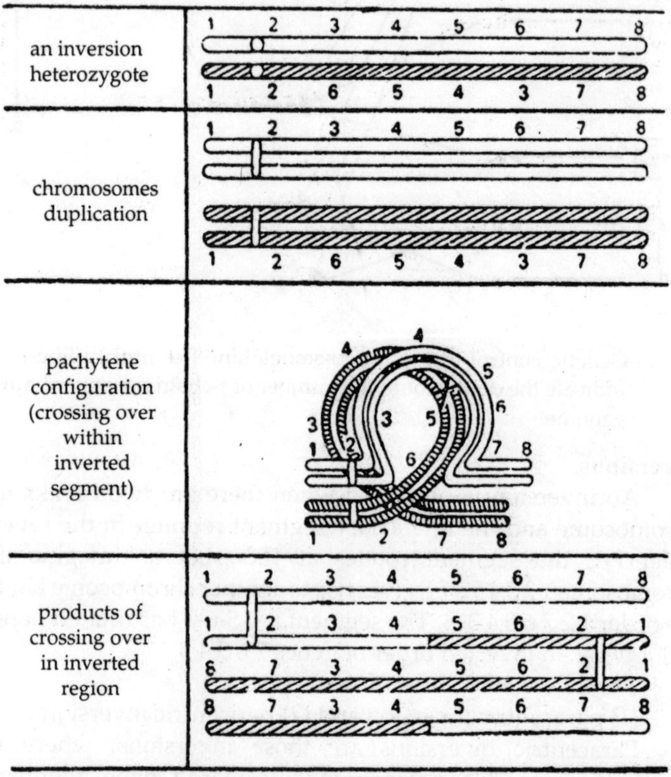

Fig. 7. Chromosome pairing and products of crossing over in a paracentric inversion heterozygote.

1. *Paracentric inversion.* A single crossing over or an odd number of crossovers in the inverted region will result into the formation of a dicentric chromosome and an acentric chromosome. Of the remaining two chromatids one will be normal and the other will carry the inversion. The dicentric chromatid and acentric chromatid will be observed at anaphase I in the form of a bridge and a fragment. Double crossovers and crossovers within and outside the inversion will give various kinds of configurations.

2. ***Pericentric inversion.*** In a pericentric inversion (where centromere is present within the inverted segment), although at pachytene, the configuration observed is similar to that described above for paracentric inversion, the products of crossing over and the configurations at subsequent stages of meiosis differ. In this case, two of the four chromatids resulting after meiosis will have deficiencies and duplications. However, unlike paracentric inversion, no dicentric bridge or acentric fragment will be observed. Consequently, an anaphase I, no bridge or fragment will be seen.

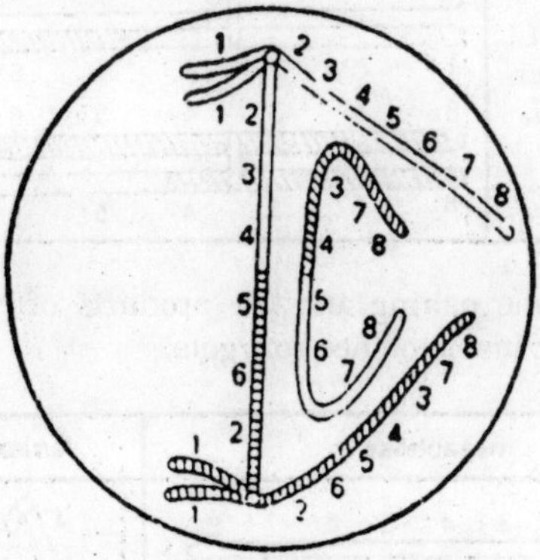

Fig. 8. Dicentric bridge and an acentric fragment at anaphase I.

However, in pericentric inversion, if the two breaks are not situated equidistant from the centromere, this will result in the change shape of the chromosome. For instance, a metacentric chromosome (with centromere in the centre) may become submetacentric and vice versa.

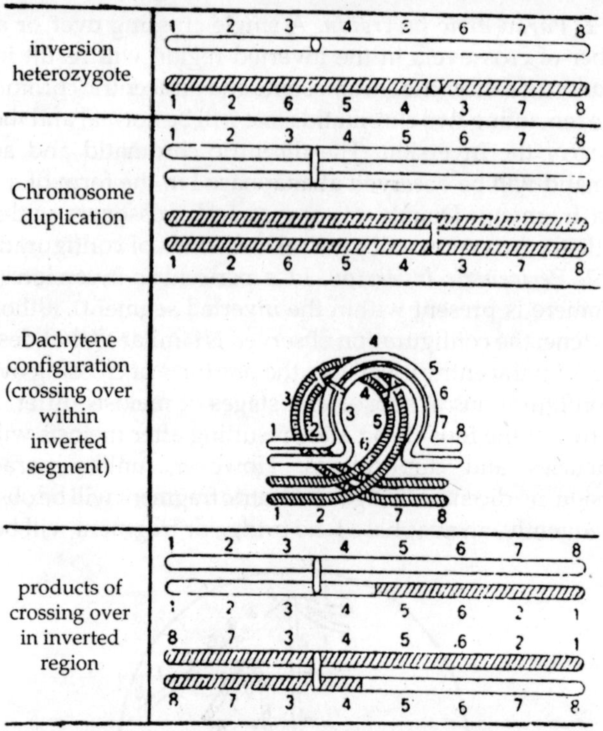

Fig. 9. Chromosome pairing and the products of crossing over in a pericentric inversion heterozygote.

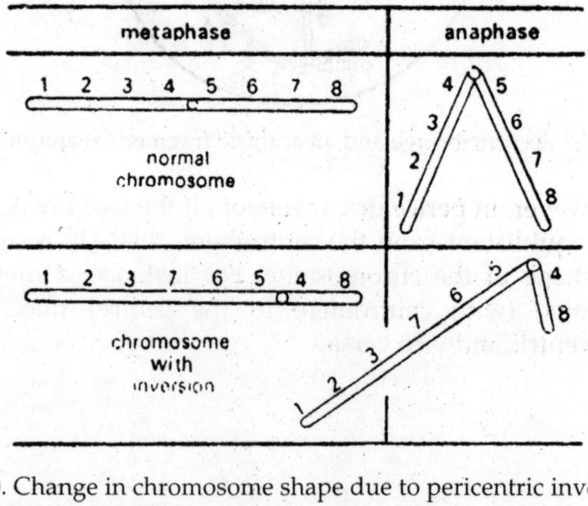

Fig. 10. Change in chromosome shape due to pericentric inversion.

Genetic Consequences of Inversion

As discussed in the preceding section on cytology, among four chromatids resulting after crossing over, the two chromatids resulting from crossing over would have deficiencies and duplications. The gametes having these chromosomes will not function. Therefore, there should be considerable gametic or zygotic lethality. In plants, there will be sufficient pollen sterility.

Since the products of single crossover will not function and the only crossover recovered will be double crossovers, the observed frequency of recombination between any two genes in question will be considerably reduced. Due to this reason, inversions, are often called crossover suppressors. This reduction in crossing over is not the actual reduction in cytological crossing over, but is the result of lack of recovery of the products of single crossovers. This property of inversion has been utilised in the production of CIB stock, used by Muller for the detection of sex linked lethal mutations.

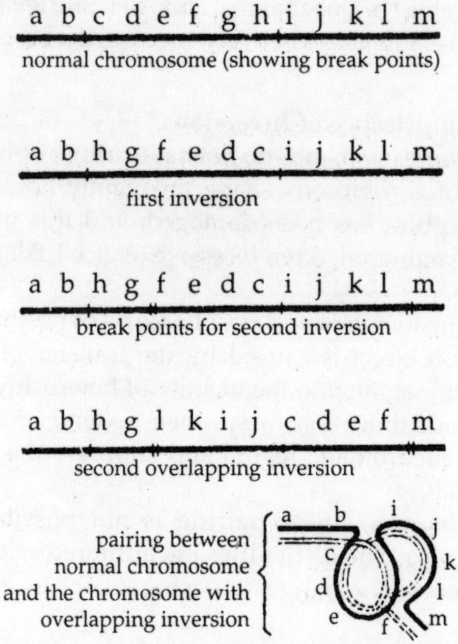

Fig. 11. Steps leading to an overlapping inversion and corresponding pachytene configuration.

Overlapping Inversions

Sometimes a second inversion takes place in a chromosome which already has one thus resulting in an overlapping inversion if the segments involved in 1st and 2nd inversions contain a common region. The gene sequences and meiotic configurations found in an inversion heterozygotes of this type.

Inversions in *Drosophila* Populations

Inversions are known to have played a significant role in evolution of different species and races of *Drosophila.* This knowledge is particularly available in *Drosophila* due to the ease of the identification of inversions in salivary gland chromosomes. They also occur in plants, but can not be so easily worked out in the absence of giant chromosomes. In *Drosophila,* however, it is obvious that inversions occurred spontaneously in nature and became established populations due to the adaptive benefit they conferred. Due adaptive value these inversions are restricted to definite localiti Th. Dobzhansky and A.H. Sturtevant even derived evolutions relationships between races on the basis of overlapping inversions.

Genetic Significance of Inversions

(i) Simple inversions do not have primary phenotypic effects other than on chromosome shape. Frequently, however, some DNA at a break point has been damaged, and this may result in an observable mutation, often recessive (e.g., c 1 B lethal mutation in *Drosophila*).

(ii) Due to inversion a peculiar kind of position effect occurs. The position effect is caused by the transfer of a gene from a euchromatic segment to the vicinity of heterochromatic segment. Heterochromatinization may then extend into a displaced, originally euchromatic region and suppress the transcription of the gene in it.

(iii) Normal linear pairing is not possible in inversion heterozygotes. The difficulties encountered with pairing cause a reduction of exchange (crossing over) in and around the inversion.

(iv) They maintain heterozygosity from generations to generations.

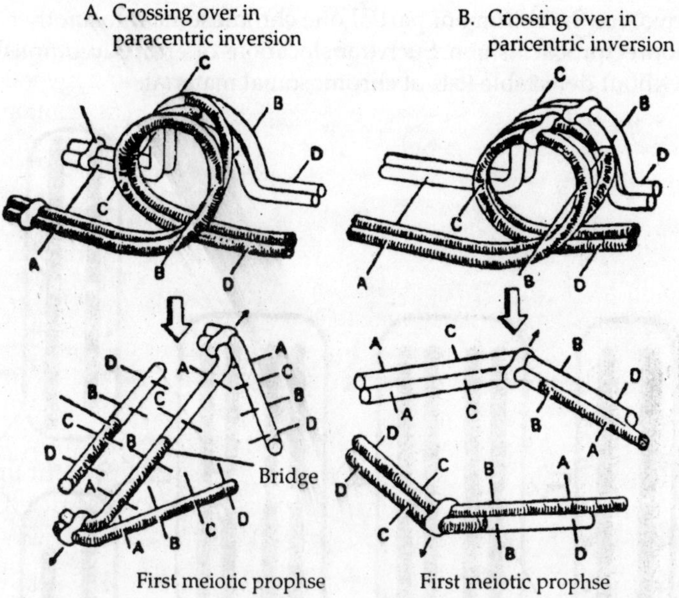

Fig. 12. Crossing over in paracentric and pericentric inversions (after SRB, Owen and Edgar, 1965).

Translocations

Translocations are chromosome rearrangements that result from the transfer of segments between two or more nonhomologous chromosomes. At least two chromosomes in the same nucleus must sustain breaks at the same time in order for such transfers to occur. Among those chromosomes with single centromeres, there are two basic types of translocations. The aneucentric type is formed when the two fragments with centromeres fuse, as do the two with no centromeres, to create a dicentric chromosome and an acentric fragment.

The rearrangements usually are not transmitted normally at cell division, so this type is not ordinarily seen except as a transitory form. However, the dicentric chromosome may in some tissues undergo a breakage-fusion-bridge cycle. Several translocations have been recognised. The type that is genetically least stable is the half-translocation in which only half of the broken ends rejoin. The unhealed segments generally are lost, and often such losses result in inviability. A second type is known as a *transposition* and involves the shifting of part of one chromosome into another in a nonreciprocal fashion. Such translocations often are accomplished without detectable loss of chromosomal material.

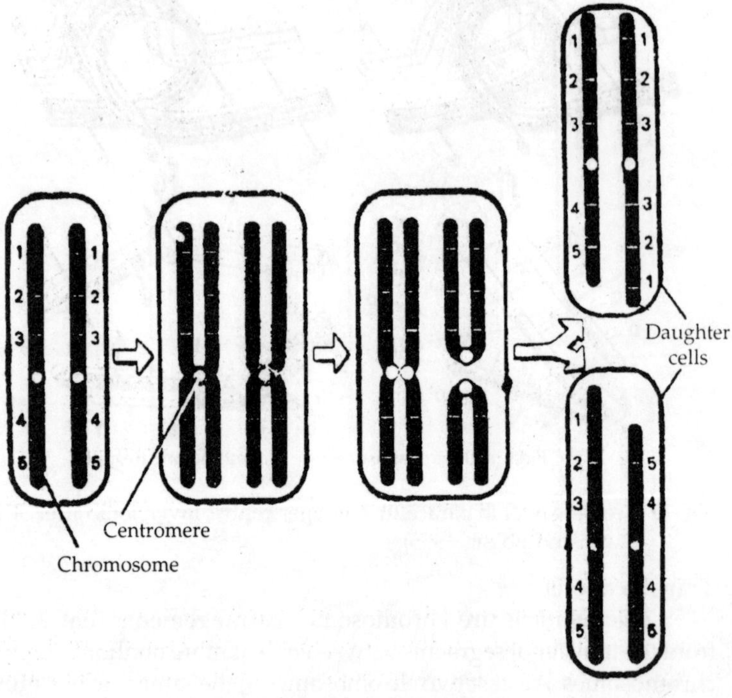

Fig. 13. Formation of a isochromosome.

Genetically the result of a transposition is to shift the linkage relationships of the transposed segment from one linkage group to another. In heterozygotes, if the transposed segment is large enough so that the chromosome missing the segment is not viable, the two linkage groups behave as a single linkage system. How this happens will become clear when the third class, the reciprocal translocation, is discussed.

Reciprocal translocations are the type most generally encountered and, like the transposition translocations, apparently can be constructed without detectable loss of genetic information. Translocation homozygotes often may be quite normal in all respects. In the homozygote, chromosome pairing and crossing

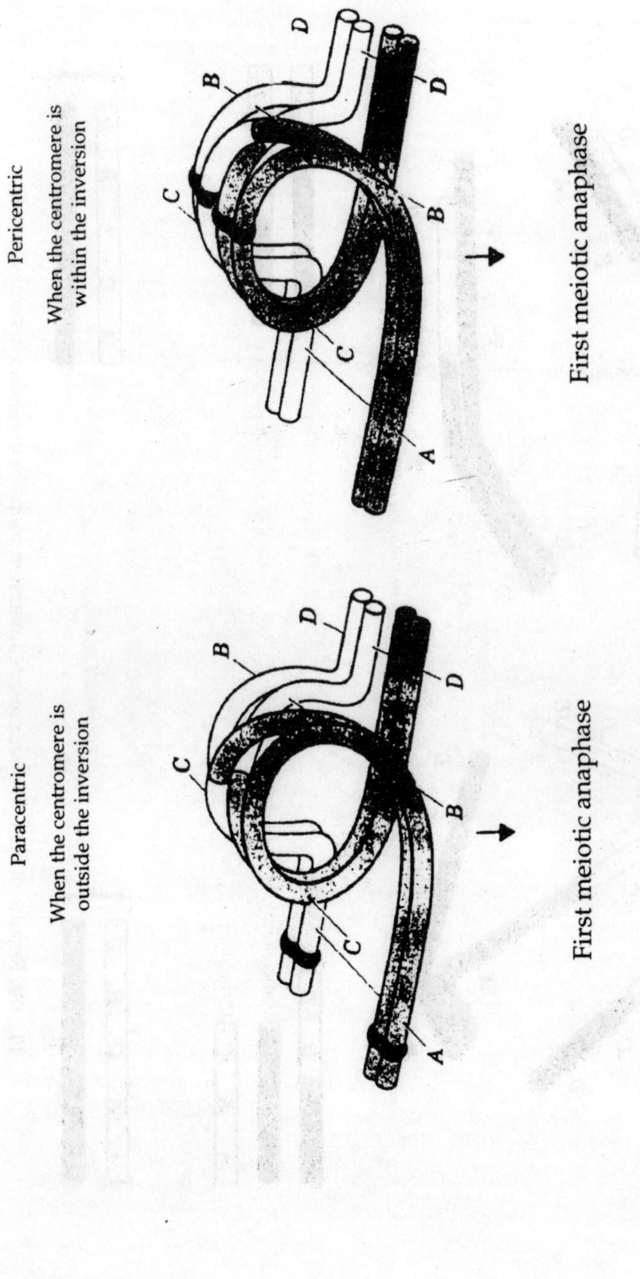
Pericentric
When the centromere is within the inversion
A
B
C
D
First meiotic anaphase
Paracentric
When the centromere is outside the inversion
A
B
C
D
First meiotic anaphase

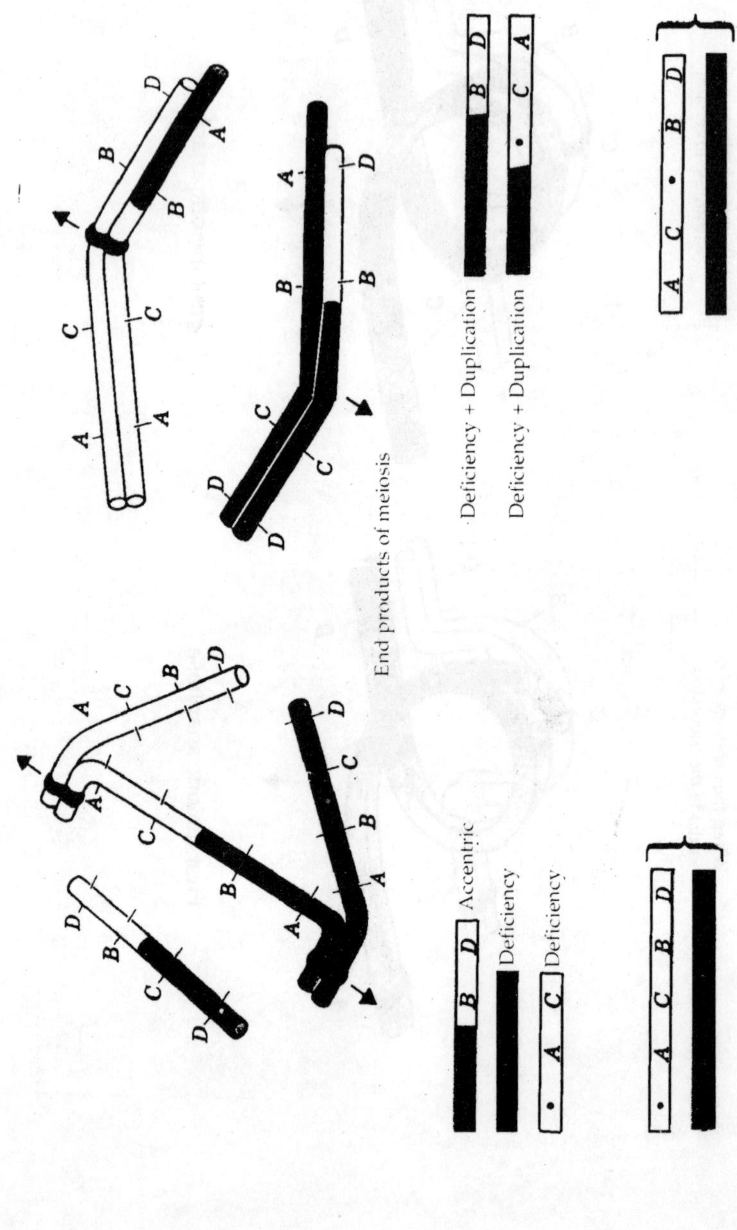

Fig. 14. Results of crossing over inside inversion loops in inversion heterozygotes.

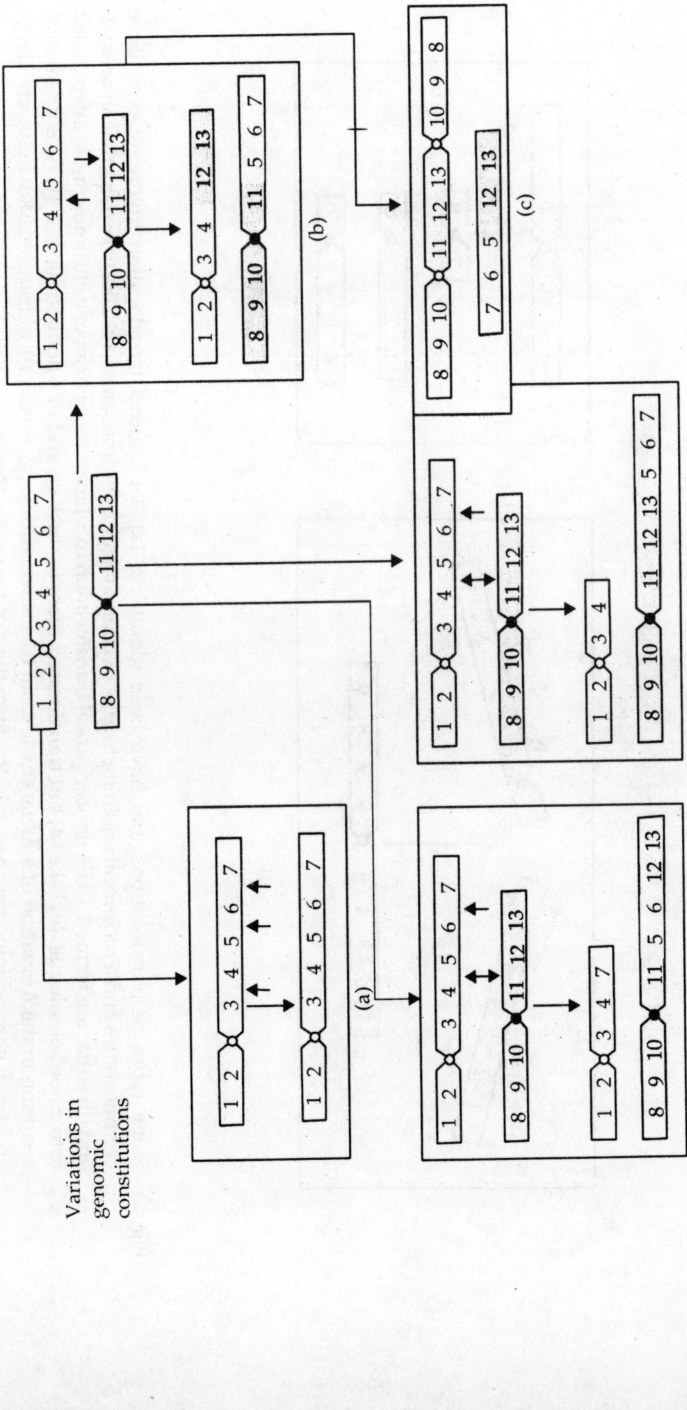
Variations in genomic constitutions
(a)
(b)
(c)
(d)
(e)

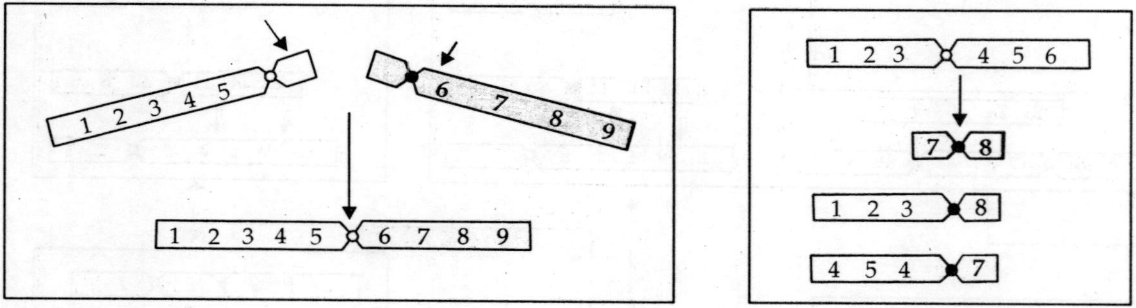

(f) (g)

Fig. 15. Some types of translocations that have been identified. (a) An internal translocation involving only three chromosomes and a reciprocal rejoining to give two monocentric chromosomes. (c) Like (b) except that an acentric and dicentric are formed. (d) A nonreciprocal translocation in which a piece from one chromosome is intercatated into a nonhomologue. (e) Like (d) but translocated piece is stuck on end of a nonhomologue. (f) A centric fusion resulting in the formation of a metacentric from two acrocentrics. (g) A reciprocal translocation between a large and small chromosome may result in the formation of two acrocentrics.

over during meiosis is normal, and, unless the size of the exchanged fragments is very different, it may not be possible to detect a homozygous reciprocal translocation cytologically unless of course there are characteristic cytological landmarks such as the banding patterns of polytene chromosomes. When translocated chromosomes are compared with their normal homologues by construction of a heterozygote, there are characteristic cytological and genetic results that are easily recognised.

Cytologically the heterozygous translocated chromosomes form a crosslike configuration as they synapse during prophase I of meiosis. As metaphase I is approached (assuming crossing over occurs), the configuration may take on a ring or double-ring arrangement as the chiasmata in each arm of the cross terminate as the homologous centromeres move apart. Genetically, because two pairs of chromosomes are involved in a single pairing complex, six different types of gametes may be produced by an organism heterozygous for a translocation. Only two of the six types will produce viable haploid offspring (as in the plant gametophyte stage) or viable zygotes when joined by a gamete carrying a normal chromosome complement. Therefore, one genetic consequence of translocation heterozygotes is partial sterility. The six different types of gametes are shown in figure. The two genetically complete types result from *alternate disjunction,* where homologous centromeres separate and the two normal chromosomes move to one pole while the two translocated chromosomes go to the other. The other four types result from *adjacent disjunction,* where one structurally normal and one translocated chromosome move to the same pole and their complementary types move to the other. This results in duplication and deficiency gametes and thus to inviability.

In organisms such as *Drosophila* where such duplication deficiency gametes are functional, it is possible to obtain (from a cross involving two parents both heterozygous for the same translocation) viable genetically complete zygotes from the combination of gametes with offsetting duplication-deficiency genotypes. A zygote produced by fertilisation of an egg with one of the type II genotypes fertilized by the other type II will have neither a duplication nor a deficiency. The same applies to the type III.

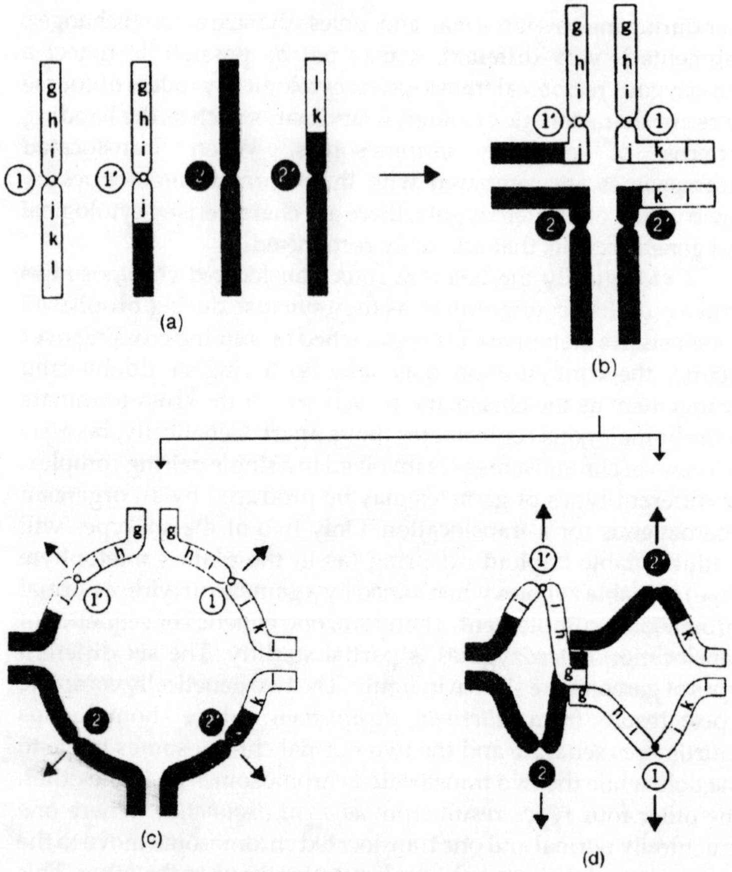

Fig. 16. Pairing in a reciprocal translocation heterozygote. (a) The genotype of the heterozygote (b) Pairing of the chromosomes in the prophase of meiosis. For simplicity the chromosomes are not shown as sister chromatics. (c) A view of the ensemble at the beginning of anaphase I. (d) Another view of the ensemble in (c).

The relative proportions of the six gamete types vary from one translocation to another with the most important variables being the length of the intervals between the translocation break points and the respective centromeres. This, of course, determines how much crossing over will occur in those intervals. As can be discerned from a study of Figure, if a crossover occurs in either or both of these designated intervals, the cells receiving the crossover chromosome following alternate disjunction now become like those produced by one or the other adjacent disjunction patterns. In a plant such as maize, where the proportion of viable to inviable products can be measured by examining pollen grains, the frequency of viable cells is usually in the neighbourhood of 50 per cent. In addition to the semisterility associated with heterozygous translocation, there is of course an alteration of linkage relationships. Translocations in natural populations appear to be more frequent in plants than in animals. Multiple translocation systems have been reported in several plant species, with one of the most elaborate occurring in the evening primrose, *Oenothera.*

In some varieties of *Oenothera,* all 14 chromosomes are involved in translocations. This results in all the chromosomes behaving as a single linkage group. These chromosome complexes are often found associated with recessive lethals, making the two homozygous conditions lethal. *O. lamarckiana,* for example, consists of two complexes, *velans* and *gaudens,* which compose a balanced lethal system since *velans · velans* and *gaudens · gaudens* homozygotes die at very early stages. At meiosis, the *velans · gaudens* combination forms a ring of 12 chromosomes and a bivalent, indicating that six of the seven pairs of chromosomes are involved in translocations.

Translocations have been detected in human beings. The best example involves chromosomes 14 and 21. This type of exchange has been associated with a condition known as Down's syndrome. It is known that Down's syndrome is caused by a trisomy for chromosome 21, and the assumption is that the translocation has caused a duplication for a part of chromosome 21, since translocation heterozygotes regularly produce duplication-deficiency gametes.

Genetic Significance of Heterozygotic Translocation

1. The heterozygous translocation produce semi-sterile organisms because between half and two third gametes fail to receive the full complements of genes required for normal development of sex.
2. Some genes which formerly assorted independently, exhibit linkage relationships after translocation has occurred; a single reciprocal translocation will reduce the number of linkage groups by one.

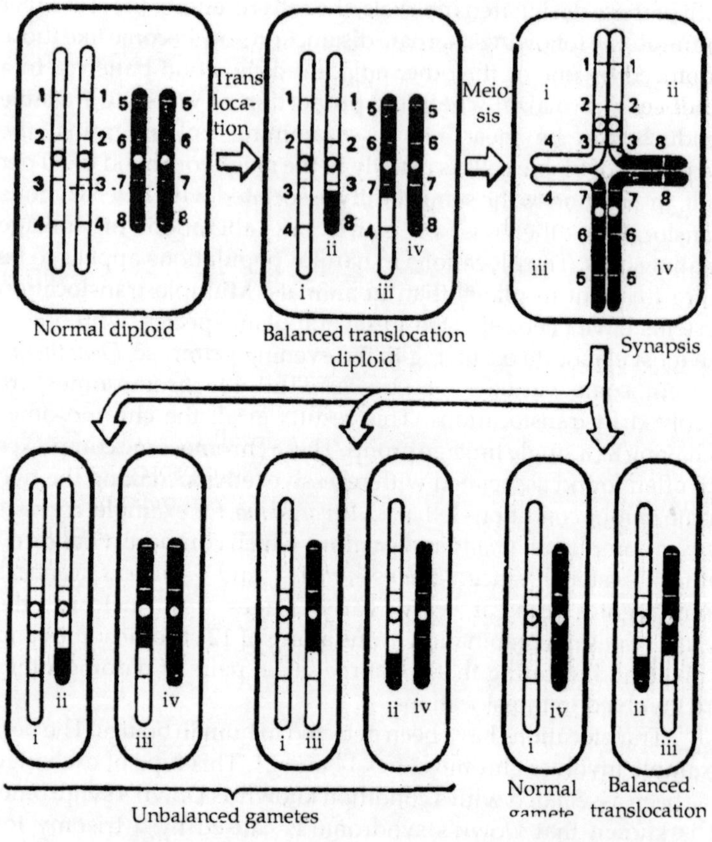

Fig. 17. Meiosis in heterozygotic translocation.

3. The phenotypic expression of a gene may be modified when it is translocated to a new position in the genome (position effect).

10

Polyploidy

Every organism has a definite number of chromosomes which is the *diploid* number (2n). During gametogenesis this number becomes half (n) and is called *haploid* number. When the number exceeds two the stage is called *polyploidy*.

In all species, individuals are occasionally found that have extra chromosomes or that lack chromosomes. These represent abnormalities in chromosome *number*. Other rare individuals are found to have an alteration in the arrangement of genes in the genome, such as by having a chromosome with a particular segment missing, reversed in orientation, or attached to a different chromosome. These represent abnormalities in chromosome *structure*. It will be seen that plants are much more tolerant of such changes than are animals, and that in animals numerical alterations often produce stronger effects on phenotype than do structural alterations.

Forms of Chromosomes

It is the centromere that determines the form or shape of the chromosome. In both mitosis and meiosis spindle fibers attach to the centromere of each chromosome during cell division and cause the sister chromatids to be separated and moved to opposite poles. Occasionally chromosomes arise having an abnormal number of centromeres. The upper chromosome has two centromeres and is said to be *dicentric*. Such aberrant chromosomes are unstable because they are frequently lost from cells when the two centromeres proceed to opposite poles during cell division; in this case the chromosome is stretched and forms a *bridge* between the daughter cells, which may not be included in either daughter nucleus or may break, with the result that each daughter nucleus receives a broken chromosome. The lower chromosome in the figure is an *acentric* chromosome, which lacks a centromere. Acentric chromosomes are also unstable because they cannot be maneuvered properly during cell division and tend to be lost. Only chromosomes that have a single centromere are regularly transmitted from parents to offspring. Chromosomes are conveniently described in terms of their form during anaphase movement. Three distinct shapes are seen, resembling a V, or a J, or an I. The shape is determined by the position of the centromere and the relative length of the lagging chromosome arms. A V-shaped chromosome has its centromere approximately in the middle, forming arms of about equal length, and is called a *metacentric* chromosome. A J-shaped chromosome has an off-center centromere forming arms of unequal length; such chromosomes are called *submetacentric.* When the centromere is very close to one end, the chromosome appears I-shaped at anaphase because the arms are grossly unequal in length; such a chromosome is termed *acrocentric.* The distinction between a metacentric, submetacentric, and acrocentric chromosome is somewhat arbitrary, but the terms are useful because they provide a physical image of the chromosome.

Chromosome evolution often tends to conserve the number of chromosome *arms* without conserving the number of *chromosomes.* For example, *Drosophila melanogaster* has two large metacentric autosomes, but many other *Drosophila* species have four acrocentric autosomes instead. Detailed comparison of the genetic maps of these species reveals that the acrocentric chromosomes in the other species correspond, arm for arm, with the large metacentrics in *D. melanogaster.* Also, chimpanzees and humans have 22 pairs of chromosomes that are morphologically similar, but chimpanzees have two pairs of acrocentrics not found in humans, and humans have one pair of metacentrics not found in chimpanzees. In this case, each arm of the human metacentric is homologous to one of the chimpanzee acrocentrics.

Variation in Chromosome Number

Every organism is characterised by two specific genomes (2n) in each somatic cell of the diploid organisms. Most animals and plants are diploids. The diploidy is related with fertility, balanced growth, great vigorosity, adaptability and survivality of the diploid organisms.

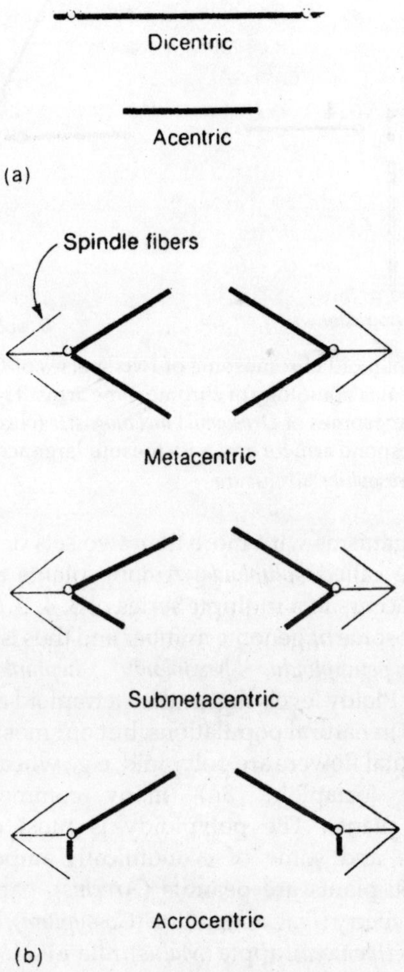

Fig. 1. (a) A diagram of dicentric (two centromeres) and an acentric (no centromere) and an acentric (no centromere) chromosome. Dicentric and acentric chromosomes are frequently lost during cell division, the former because the two centromeres, may bridge between the daughter cells, and the latter because the chromosome cannot attach to the spindle fibers. (b) The three possible shapes of chromosomes in anaphase. The centromeres are shown in red.

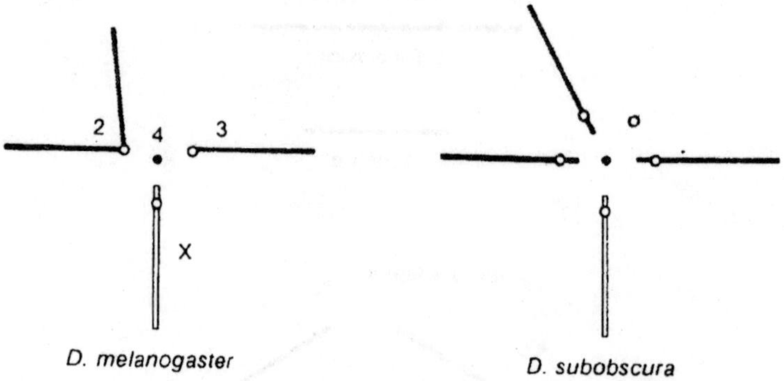

Fig. 2. The haploid chromosome of two species of *Drosophila*. Shading indicates homology of chromosome arms. The large metacentric chromosomes of *Drosophila melanogaster* (chromosomes 2 and 3) correspond arm for arm with the four large acrocentric autosomes of *Drosophila subobscura*.

The organisms with more than two sets of chromosomes or genomes are called *polyploids*. Among plants and animals, the polyploidy occurs in a multiple series of 3, 4, 5, 6, 7, 8, etc., of the basic chromosome or genome number and thus is causing *triploidy, tetraploidy, pentaploidy, hexaploidy, heptaploidy, octaploidy,* respectively. Ploidy levels higher than tetraploid are not commonly encountered in natural populations, but our most important crops and ornamental flowers are polyploid, e.g., wheat (hexaploid, 6n), strawberries (octaploid, 8n), many commercial fruit and ornamental plants. The polyploidy is most common among angiosperms and some of economically important polyploid angisopermic plants are peanuts (*Arachis*), oats (*Avena*), coffee (*Coffea*), strawberry (*Fragaria*), cotton (*Gossypium*), barely (*Hordeum*), sweet potato (*Ipomoea*), apple (*Malus*), alfa-alfa (*Medicago*), banana (*Musa*), tobacco (*Nicotina*), plum (*Prunus*), sugar cane (*Saccharum*), potato (*Solanum*), sorghum (*Sorghum*), clover (*Trifolium*), and wheat (*Triticum*).

A continuous polyploid series has been reported in rose plant. A euploid series of basic number of 7 (monoploid) including diploids (2n = 14), triploids (21), tetraploids (28), pentaploids (35), hexaploid (42), and octaploid (56) has been reported in different species of *Rosa*. Likewise, the genus *Chrysanthemum* has basic chromosome number 9 and has a euploidic series of diploid

(2n = 18), tetraploids (4n = 36), hexaploids (6n = 54), octaploids (8n = 72) and decaploids (10n = 90) in its different species. The genus *Solanum* has basic chromosome number 12 and has a euploidic series of diploids (2n = 24), triploids (3n = 36), tetraploids (4n = 48), pentaploids (5n = 60), hexaploids (6n = 72), octaploids (8n = 96), and decaploids (10n = 120) in its different species.

Table 1
Some Common Polyploid Plants

Genus and Species	*Number of chromosome*	*Ploidy*
1. *Sorghum versicolour*	5	Monoploid ($n = 5$)
S. vulgare	10	Diploid ($2n = 2 \times 5$)
S. halepense	20	Tetraploid ($4n = 4 \times 5$)
2. *Triticum monococcum*	14	Diploid ($2n = 2 \times 7$)
T. diococcum	28	Tetraploid ($4n = 4 \times 7$)
T. aestivum	42	Hexaploid ($6n = 6 \times 7$)
3. *Gossypium arboreum*	13	Monoploid ($n = 13$)
G. hirsutum	26	Diploid ($2n = 2 \times 13$)
4. *Trifolium hybridum*	8	Monoploid ($n = 8$)
T. repens	16	Diploid ($2n = 2 \times 8$)
T. medium	40	Pentaploid ($5n = 5 \times 8$)

Polyploidy is rare in animals, however, has been reported in *Ascaris, Daphnia, Cypris, Drosophila, Bombyx mori, Habrobracon*, bees, wasps, *Artemia, Echinus, Rana esculenta, Rana pipiens, Ambystoma, Jeffersoniarium, Cricetus cricetus* (hamaster), etc.

Origin of Polyploidy

Polyploidy can be originated by two basic irregular processes from which polyploids may evolve from diploid plants and become established in nature:

1. Somatic doubling—cells sometimes undergo irregularities at mitosis and give rise to meristematic cells that perpetuate these irregularities in new generations of plants.

2. Reproductive cells may have an irregular reduction or equation division in which the sets of chromosomes fail to separate completely to the poles at anaphase. Both sets thus become incorporated in the same nucleus resulting in the doubling of chromosome number in the gamete. Thus, a triploid originates by the fusion of a haploid gamete (n) with a diploid gamete (2n), the later of which may be originated by irregularities during meiosis. Likewise, a tetraploid may be originated by the somatic doubling of the chromosome number or by union of unreduced diploid gametes. The somatic doubling of genome is accomplished either spontaneously or it can be induced in high frequency by exposure to chemicals such as **colchicine,** etc., or heat or cold. Other levels of polyploidy are also originated by same methods.

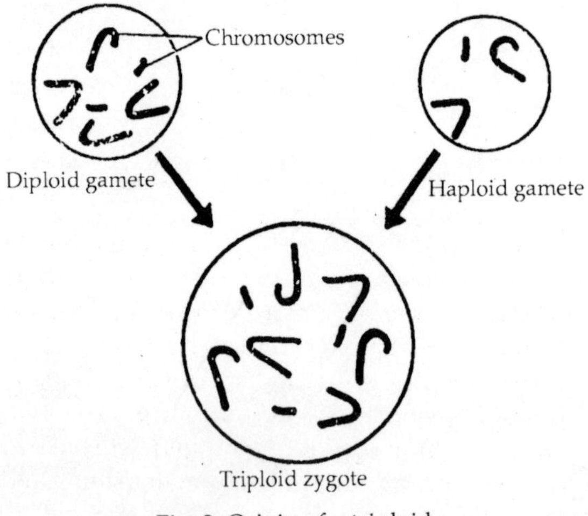

Fig. 3. Origin of a triploid.

Induced Polyploidy

The polyploidy can be induced in common diploid organisms by following methods:

Cell generation. In certain bryophytes such as mosses, the polyploidy has been induced by cutting their diploid sporophytes and keeping the sporophytes in moist conditions. The cells of the cut ends regenerated threads which were true protonema and produced diploid gametophytic generation instead of monoploidic generation.

Physical agents. Following kinds of physical conditions induce polyploidy in plants:

(i) Temperature shocks. Extreme temperature changes sometimes result in a higher frequency of polyploid cells.

(ii) Centrifugation. The centrifugation of seedlings of plants causes polyploidy in their cells. In *Nicotiana*, polyploidy has been induced by this method.

(iii) X-rays. The radioactive substances such as radium and X-rays have been found to induce polyploidy in normal diploid plant cells.

Chemical agents. Chemicals like **colchicine,** chloral hydrate, acenaphthene, veratrine, sulfanil amide, ethyl, mercury chloride, hexachlorocyclohexane have been reported to induce polyploidy in plants. These chemical substances when applied to dividing diploid cells, they disturb the mitotic spindle and cause non-segregation of already duplicated chromosomes and thus, convert the diploid cells into tetraploid cells. The tetraploid cells, likewise, are converted into different levels of polyploidy.

Kinds of Polyploidy

On the basis of the source of chromosomes is of two types of polyploidy have been distinguished.

1. Autopolyploidy. The prefix "auto" indicates that the ploidy involves only homologous chromosome sets. Somatic doubling of a diploid produces four sets of genomes of a tetraploid and likewise, somatic doubling of a tetraploid produces eight sets of genomes of a octaploid. For example, if a diploid species has two similar sets of chromosomes or genomes (AA), on autotriploid will have tree similar genomes (AAA), and an autotetraploid, will have four such genomes (AAAA). Since an autotriploid remains sterile and cannot produce seeds, therefore, it has great commercial value in producing seedless varieties of economical plants. For example, in Japan, *H. Kihara* produced seedless watermelons, which were autotriploids. Common "doob' grass of UP and Bihar is an autotriploid. Other common seedless autopolyploids are grapes, sugarbeet, banana, etc. In *O. lamarckiana*, the giant mutant described by **de Vries** was later on discovered to be an autotetraploid. Further, whenever autopolyploids, originate in nature, these would be eliminated due to natural selection.

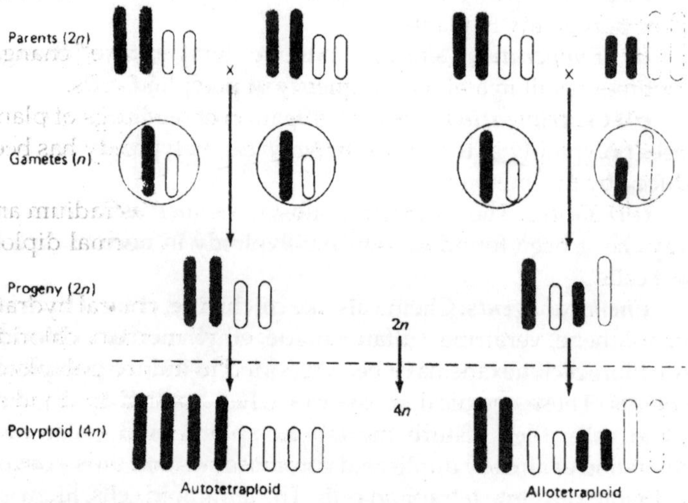

Fig. 4. Origin of the genomes in an autotetraploid versus an allotetraploid.

2. ***Allopolyploids.*** Polyploidy may also result from the doubling of chromosome number in a F_1 hybrid which is derived from two distinctly different species. This will bring two different sets of chromosomes in F_1 hybrid. The number of chromosomes in each of these two sets may differ. Let A represent a set of chromosomes (genome) in species X, and let B represent another genome in a species Y. The F_1 will then have one A genome and another B genome. The doubling of chromosomes in this F_1 hybrid (AB) will give rise to a tetraploid with two A and two B genomes. Such a polyploid is called an allopolyploid or amphidiploid.

(a) ***Raphanobrassica*** is a classical example of allopolyploidy. In 1927, G.D. Karpechenko, a Russian geneticist, reported a cross

between *Raphanus sativus* (2n = 18) and *Brassica cleracea* (2n = 18) to produce F_1 hybrid which was completely sterile. This sterility was due to lack of chromosome pairing, since there is no homology between the genomes from *Raphanus sativus* and *Brassica oleracea.* Among these sterile F_1 hybrids, Karpechenko found certain fertile plants. On cytological examination these fertile plants were found to have 2n = 36 chromosomes, which showed normal pairing into 18 bivalents.

(b) Evolution of wheat. Common cultivated wheat is another important example of allopolyploidy, although its allopolyploid nature has now been questioned. There are three different chromosome numbers in the genus *Triticum,* namely 2n = 14, 2n = 28 and 2n = 42. The common wheat is a hexaploid with 2n = 42 and is derived from three diploid species: (i) AA = *Triticum aegilopoides* (2n = 14)? (ii) BB = *Aegilops speltoides,* (2n = 14) and (iii) DD = *Aegilops squarrosa* (2n = 14). The hexaploid wheat, therefore, is designated as AABBDD; the tetraploid (2n = 28) as AABB and diploid (2n = 14) as AA. There is, however, evidence available now which suggests that A, B and D genomes from three diploid species mentioned above are not much different from one another, so that it is believed that the three diploid progenitors of common hexaploid wheat were derived from a common ancestor. For this reason, the common hexaploid wheat is now considered an autopolyploid rather than an allopolyploid. At best, it may be segmental allopolyploid.

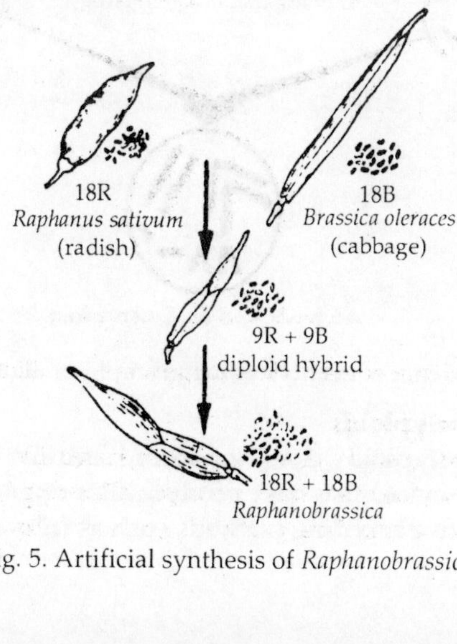

Fig. 5. Artificial synthesis of *Raphanobrassica.*

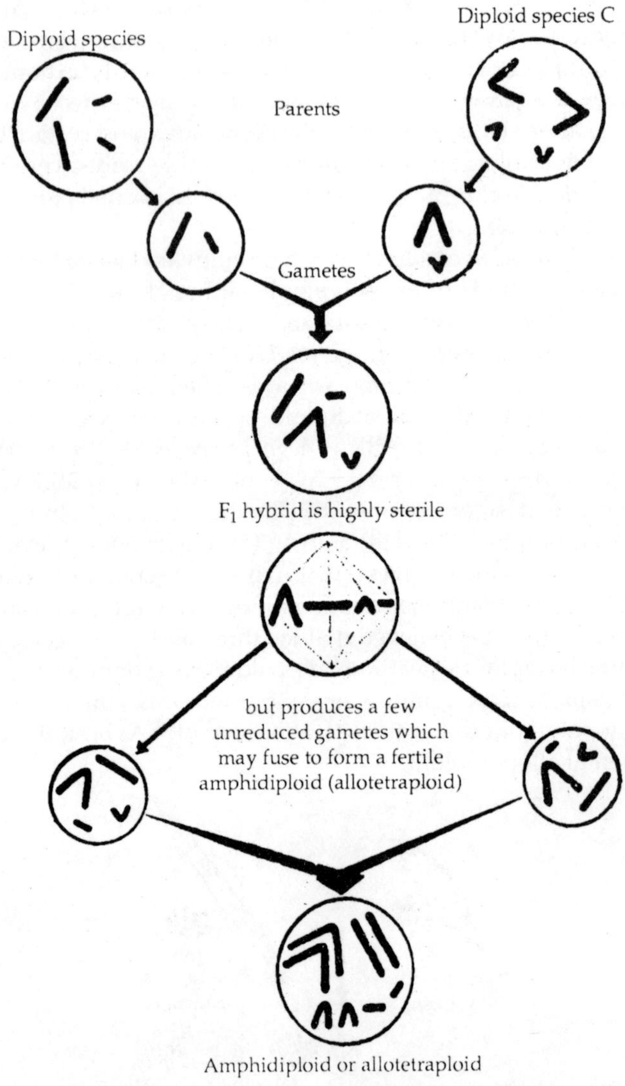

Fig. 6. A sequence of events leading to amphi-or allotetraploidy.

Induced Allopolyploids

The allopolyploidy does not occur naturally in the nature, however, some cytologists have produce allopolyploidy in certain plants by selective breeding methods such as follows:

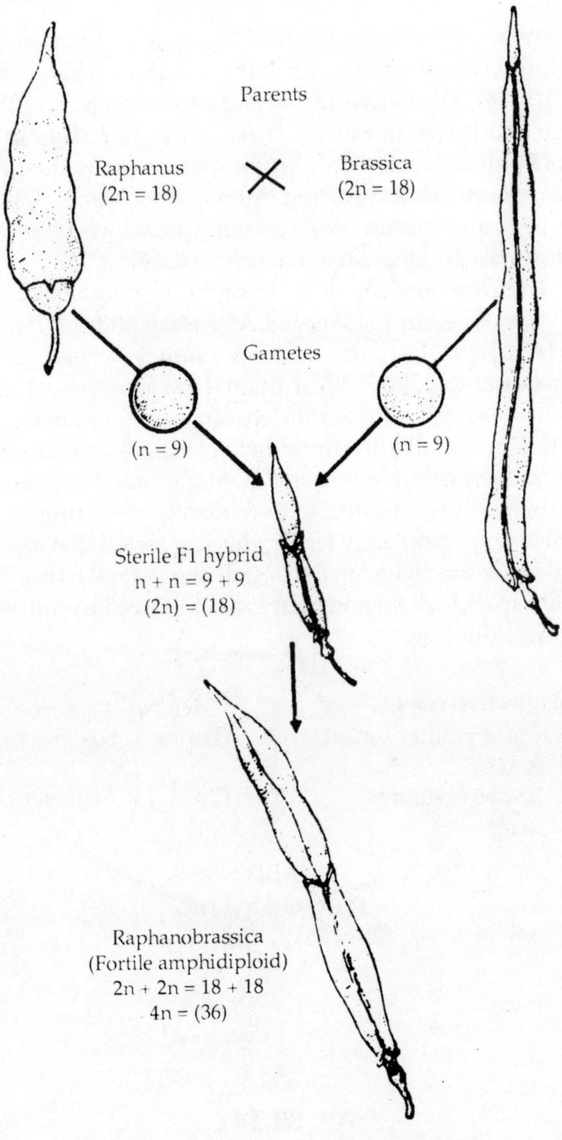

Fig. 7. The origin of *Raphanobrassica*.

(a) The Russian cytologist, G.D. Karpechenko (1928) first of all produced allotetraploid genus called *Rhaphanobrassica* from the

artificial crosses between vegetables belonging to different genera, the radish (*Raphanus sativum,* 2n = 18) and the cabbage (*Brassica oleracea,* 2n = 18). The F_1 hybrids were diploid and having the root of cabbage and leaves of radish. These plants were highly sterile because of failure of each set of chromosomes to provide sufficient genetic homology to affect pairing. Among these sterile F_1 hybrids, however, he found certain fertile allotetraploids which contained 36 chromosomes and were named as *Rhaphanobrassica.*

(b) Triticum spelta, is a hexaploid wheat which was artificially produced in 1946 by *E.S. McFadden* and *E.R. Sears* and also by *H. Kihara.* They crossed an emmer wheat, Triticum *dicoccoides* (tetraploid: 2n = 28) with goat grass, *Aegilops squarrosa* (diploid; 2n = 14) and doubled the chromosome number in the F_1 hybrids. This artificially produced hexaploid wheat was found to be similar to the primitive wheat *T. spelta.* When the synthesized hexaploid wheat was crossed with naturally occurring *T. spelta,* the F_1 hybrid was completely fertile, this suggested that hexaploid wheat must have originated in the past due to natural hybridization between tetraploid wheat and goat grass followed by subsequent chromosome doubling.

P_1: *Triticum dicoccoides* × *Aegilops squarrosa*

(Tetraploid emmer wheat) (Diploid goat grass)

AABB DD

(2n = 28; 14 bivalents) (2n = 14; 7 bivalents)

↓

F_1 ABD

(Triploid hybrid)

(2n = 21; 21 univalents)

↓ Colchicine

AA BB DD

Synthesized hexaploid wheat

(Triticum spelta)

(2n = 42; 21 bivalents)

Fig. 8. Artificial production of the allohexaploid wheat.

(c) One more interesting case of allotetraploidy has been observed in the production of a rust resistant allotetraploid wheat plant. Common wheat plant, *Triticum vulgare* is susceptible to leaf rust, a seriouse disease caused by the fungus *Puccinia triticina.* A wild grass of the Mediterranean region, *Aegilops umbellulata* is completely resistant to this disease. Sears (1956) have transferred the genes of rust resistance of *A. umbellulata* into *T. vulgare* genome by following method: He crossed the plants of *A. umbellulata* with *T. dicoccoides* and got sterile hybrid which by treatment with colchicine was transformed into a rust-resistant, fertile allotetraploid having 21 pairs of chromosomes. The allotetraploid was crossed to *T. vulgare* and a fertile, rust-resistant hybrid was produced.

(d) Gossypium hirsutum. It is a New world cotton plant, it is another excellent example of allopolyploidy. Old world cotton, *Gossypium herbaceum,* has 13 pairs of chromosomes, while American or "upland cotton" also contains 13 pairs of chromosomes, J.O. Beasley crossed the old world and American cottons and doubled the chromosome number in the F_1 bybrids. The allopolyploids thus produced resembled the cultivated New world cotton and when crossed with it gave fertile F_1 hybrids. These results, thus, suggested that tetraploid *Gossypium hirsutum* originated from two diploid species, namely *G. herbaceum* (2n = 26) and *G. raimondii* (2n = 26).

Gossypium herbaceum (Old world cotton; 2n = 26; 13 bivalents)	×	*Gossypium raimondil* (American or Upland) 2n = 26; 13 bivalents)

↓

F_1 hybrid
(2n = 26; 26 univalents)

↓ Colchicine

New world cotton
(Gossypium hirsutum)
(2n = 52; 26 bivalents)

Fig. 9. Artificial synthesis of New world cotton.

Effects of Polyploidy on the Organism

The stage polyploidy has the following effects on the organisms:

Genetical effects of polyploidy. The genetical effects of polyploidy are as given below:

(i) Segregation in polyploids: The polyploidy often brings about sterility. For example, the extra set of chromosomes of the triploid is distributed in various combinations to the germ cells, resulting genetically unbalanced gametes. The autotetraploids however, are fertile because in them segregation of chromosomes is complete.

Segregation in autotetraploids. In a autotetraploid stage every nucleus has genes in quadruplicate, therefore, the genetics of a tetraploid becomes comparatively more complicated than a diploid. For example, if a given gene exists in two allelic forms *A* and *a,* then five genotypic combinations can be formed: *AAAA (quadruplex), AA Aa (Triplex), AA aa (duplex), Aa aa (simplex)* and *aa aa* (nulliplex). If allele *A* is dominant to allele *a,* the allele *a* will manifest itself only in the nulliplex condition.

Thus, A duplex tetraploid may produce *AA, Aa* and *aa* gamets. in a 1:4:1 ratio. When two duplex tetraploids are crossed, the resulted progeny will be such as follows:

Female gametes → / Male gametes ↓	1AA	4Aa	1aa
1AA	AA AA 1	AA Aa 4	AA aa 4
4Aa	AA Aa 4	AA aa 16	Aa aa 4
1aa	AA aa 1	Aa aa 4	aa aa 1

Thus, such cross will produce the progeny in following genotypic ratios: 1 quadruplex: 8 triplex: 18 duplex: 8 simplex: 1 nulliplex. The phenotypic ratio will be 35 dominant and one recessive.

(ii) Xenia. When one allele in the pollen is able to mask the effect of double dose of another allele in the resulting endosperm, the former is said to exhibit *xenia* over the latter. For example, in corn, starchy endosperm is governed by a gene S which shows xenia with respect to its allele, for sugary endosperm (s). Four genotypes are possible for these triploid cells: starchy = SSS, SSs, Sss, sugary = sss.

Polyploids offer an opportunity for studying *dosage effects*, *i.e.*, how two or more alleles of one locus behave in the presence of a single dose of an alternative allele. Dominance refers to the masking effect which one allele has over another allele.

Phenotypic Effects of Polyploidy

In polyploidy the multiple genomes show the following phenotypic effects:

(i) Tallness. The most usual phenotypic effect of *polyploidy* is tallness of the plant. The tetraploid plants, for example, have large sized pollen grains, cells of leaves, stomata, xylem, etc. than a diploid plant. They are more huskier and vigorous than diploids. The phenomenon of gigantism has various economic significances because due to it large-sized fruits, seeds and flowers are obtained from the economically important plants.

(ii) Physiological effect. The ascorbic acid contents have been reported to be higher in tetraploid cabbage and tomato than in corresponding diploid species. Further, corn meal of a tetraploid maize seed contains 40 per cent more vitamin A than corn meal from a diploid.

Segmental allopolyploids. In some allopolyploids, the different genomes which are present are not quite different from one another. Consequently in these polyploids chromosomes from different genomes do pair together to some extent and multivalents are formed. This means that segments of chromosomes and not the whole chromosomes are homologous. Therefore, such allopolyploids are called segmental allopolyploids according to Stebbins (1943, 1950). These segmental allopolyploids are intermediate between autopolyploids and allopolyploids and can be identified by their meiotic behaviour.

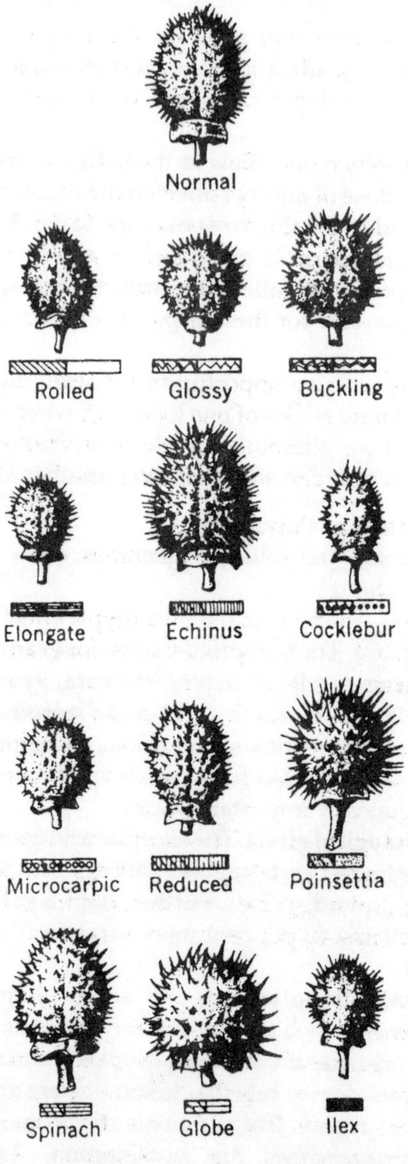

Fig. 10. Chromosomal mutants in the Jimson weed (Datura). The capsule of the normal plant is shown at the top, and the results of the duplication of each of the twelve chromosomes of this plant are shown by the other twelve seed capsules.

It is also believed that most of the naturally occurring polyploids are neither true autopolyploids nor true allopolyploids but are rather segmental allopolyploids. Our common hexaploid bread wheat is also regarded as a segmental allopolyploid, because the three diploid genomes (A, B and D) are related (homologous) to each other.

Aneuploidy

Variations in chromosome number may occur which do not involve whole sets of chromosomes (genomes) but only parts of a set (genome) are called *aneuploidies* (Gr., *aneu*-uneven; ploid-unit). The aneuploidy may be of following types:

*1. **Monosomy.*** The diploid organisms which lack in one chromosome of a single homologous pair of chromosome are called *monosomics* with the genomic formula $2n-1$. A monosomic produces two kinds of gametes, (n) and ($n-1$), because, the single chromosome without a pairing partner may go to either pole during is meiosis. In plants $n-1$ gametes remain non-functional, but, in animals result in genetic imbalance which is manifested by high mortality or reduced fertility.

2. ***Nullosomy.*** The diploid organisms which have lost one pair of homologous chromosomes are called *nullosomics* with the genomic formula, $2n-2$. The nullosomics exhibit reduced vigour, fertility and survivality, but, polyploidic nullosomics such as nullosomic hexaploid wheat ($6n-2$) survive to maturity because of the genetic redundancy in polyploids. In human being there may be 22 possibilities of nullosomics.

3. ***Trisomy.*** The diploid organisms which have one extra chromosome are called *trisomics.* They have the chromosomal formula $2n+1$. In a trisomic, one of the pairs of chromosomes has an extra member, therefore, forms a trivalent structure during meiosis. During anaphase of meiosis, two chromosomes go to one pole and one chromosome to another pole and thus, two types of gametes $n+1$ and n are resulted. The trisomy has variable effects on the phenotype of the organism and in man trisomy of autosome 21 cause mongolism. In human being there may be 23 possibilities of trisomy.

Production of Trisomics

Trisomics may originate spontaneously due to production of $n+1$ type of gametes rarely as a result of non-disjunction of a bivalent. However, more often trisomics are produced artificially either by selfing the triploids (produced by crossing diploid and autotetraploids) or by crossing these triploids as females with diploids as male ($3x \times 2x$). In either case trisomics are obtained in large number and can be identified through the phenotypic effects of individual chromosomes.

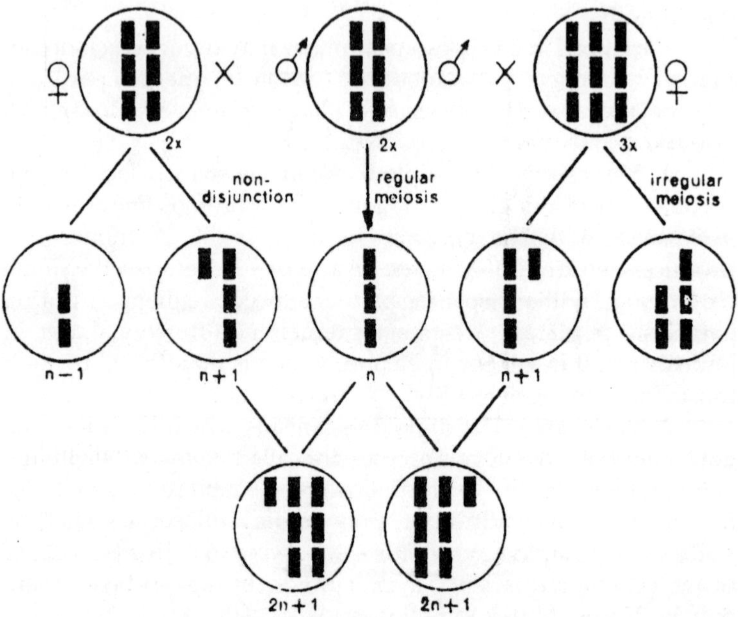

Fig. 11. Production of trisomics due to formation of $n + 1$ type of gametes in diploid ($2x$) and triploid ($3x$) individuals.

Trisomy-21 leads to the nonfamilial pattern of Down syndrome in human beings. The risk of a woman giving birth to a child with this disorder increases with increasing age of the mother. The increase in risk is presumably due at least in part to the increasing age of the oocyte, which was produced long before the woman herself was born. As she ages, the oocytes encounter a variety of changing internal environments. These physiological factors may be responsible for higher probability of meiotic nondisjunction when the oocyte matures.

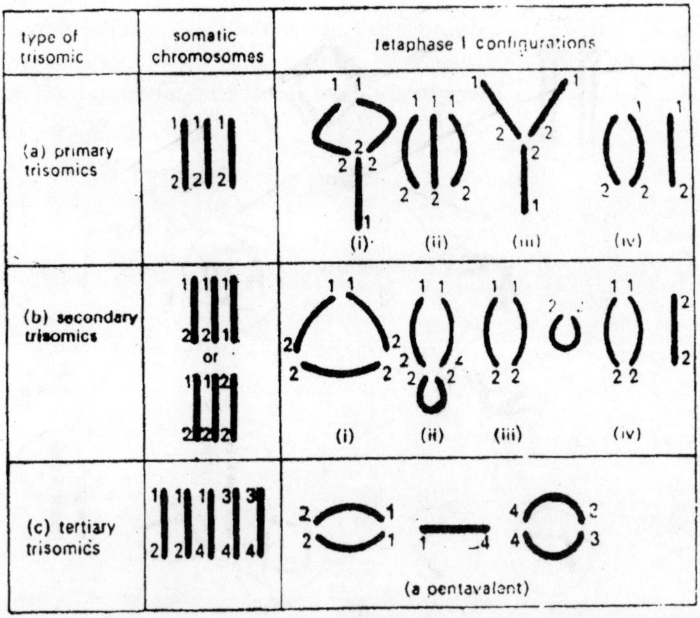

Fig. 12. Different types of trisomics and their meiotic configurations at metaphase I.

Since trisomy-21 is an autosomal disorder, it occurs equally often in both sexes. It is characterized by a number of physical abnormalities as well as mental retardation, which may be very severe in some cases. The average life expectancy has been lengthened with improved medical care, but relatively few individuals survive into their twenties.

If an older woman wishes to know whether her pregnancy will lead to the birth of a baby with Down syndrome, she may be informed specifically after **amniocentesis.** After a sample of amniotic fluid has been removed from the uterus by the physician, the fetal cells that are shed normally into this fluid can be examined. The presence of an extra chromosome 21 in a nucleus that has 47 chromosomes is direct evidence that the child will be born with the disorder.

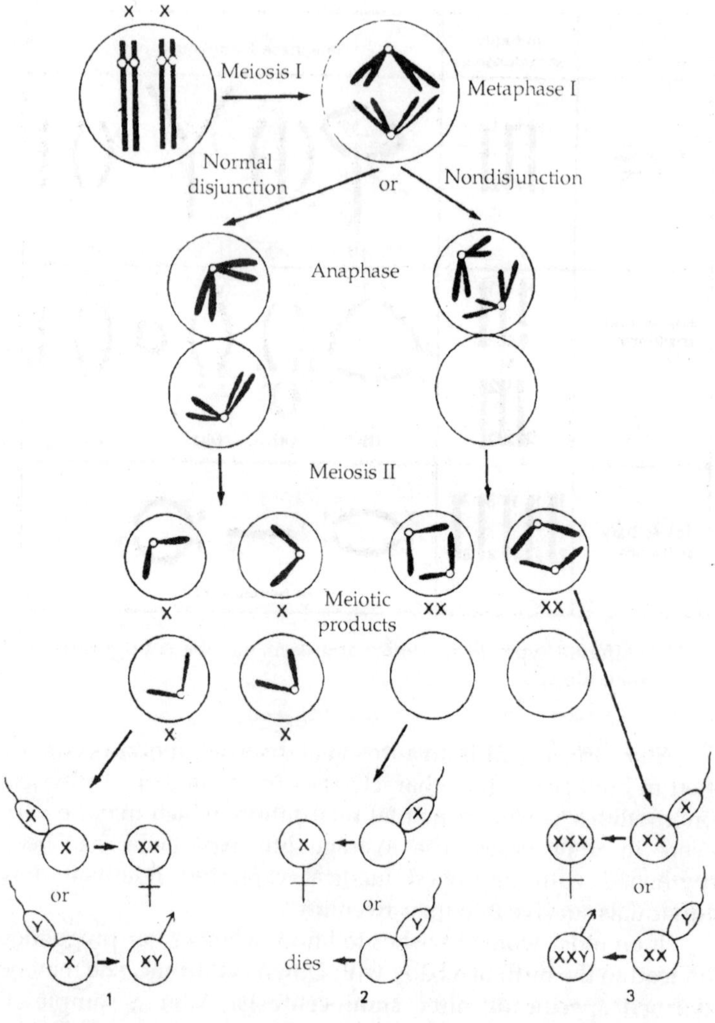

Fig. 13. Nondisjunction at meiosis can lead to gametes with too few or too many chromosomes and to aneuploid individuals produced by fusions involving such gametes. Nondisjunctions may involve sex chromosomes, as shown here, or any of the autosomes.

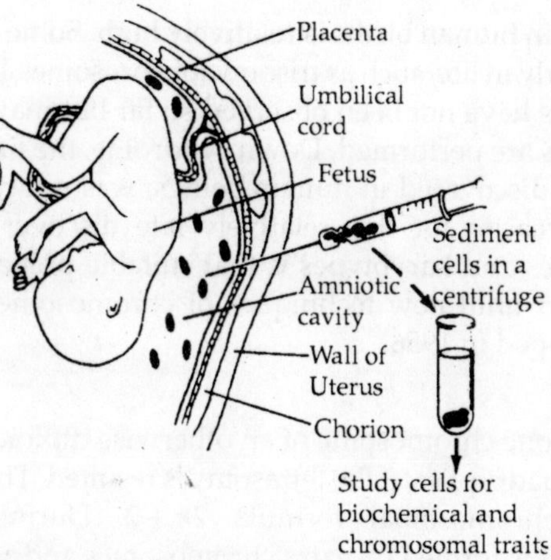

Fig. 14. Amniocentesis. A sample of amniotic fluid is withdrawn from the uterus and is processed for cytological and biochemical analysis of fetal traits.

Table 2
Frequency of Selected Aneuploidies in Humans

Aneuploidy	*Frequency of live born*
sex chromosome anomalies	
Turner syndrome (45, X)	1/2500 females
triplo-X (47, XXX)	1/1250 females
tetra-X (48, XXXX)	very low (about 20 cases known)
Klinefelter syndrome (all types)	1/800 males
XYY male	1/900 males
autosomal anomalies	
trisomy-13	1/4,000 to 10,000
trisomy-18	1/8,000
trisomy-21	1/700

The prospective parents may then make preparations for this birth or elect to have the fetus aborted. The decision is theirs alone, once they have been apprised of the situation. The incidence of aneuploidy in human births is relatively high. Some aneuploidies are lethal early in life, such as trisomy for chrosomes 13 or 18. Some aneuploidies have not been observed so far but may be found as more studies are performed. Down syndrome, the first autosomal aneuploidy discovered in human beings, was not reported until 1959. One reason for this relatively late discovery on human chromosomes and karyotypes is that suitable preparations were not achieved until new techniques of chromosome preparation were developed in 1956.

Tetrasomy

When one chromosome of an otherwise diploid organism is present in quadruplicate, the tetrasomy is resulted. The tetrasomics have the chromosomal formula $2n + 2$. During meiosis a quadrivalent is formed by extra chromosomes and segregation of chromosomes occurs like autotetraploids. In human being there may be 23 possibilities of tetrasomy.

Aneuploidy may also arise without gain or loss of chromatin by the process of **centric fusion,** in which two smaller chromosomes fuse to form one larger one. Usually, two smaller acrocentrics fuse to produce a large metacentric or submetacentric chromosome.

It was originally believed that one of the two centromeres was lost during centric fusion and that the larger derivative had only a single centromere. Electron microscopy revealed, however, that two centromere structures are contained in the single centromere region of a chromosome known to be derived by centric fusion. Some investigators suspected these results on the basis of their observations of various invertebrate species. In some of these populations, individuals had different chromosome *numbers* but they had the same number of chromosome *arms* in a genome. It has since been found that chromosomes which do undergo centric fusion, reducing the chromosome number but not the number of arms, may also dissociate into acrocentrics.

The chromosome number is increased in such individuals but the number of chromosome arms remains unchanged. Centromere retention explains these reversible changes since new centromeres cannot be created from other parts of a chromosome. Neither the mechanism for centric fusion nor the mechanism for dissociation is known. Centric fusion has played an important role in genome evolution in many animal groups

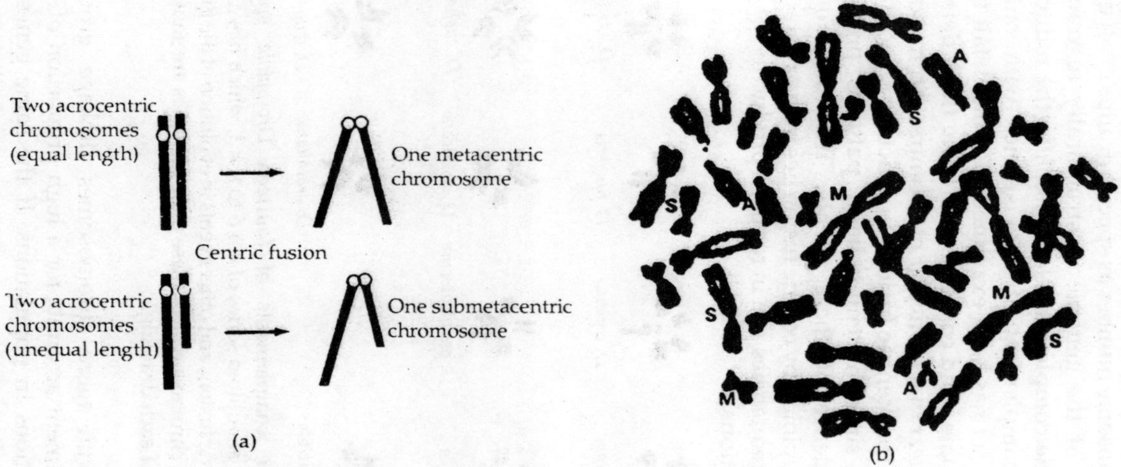

Fig. 15. Centromere location in chromosomes. (a) Centric fusion may lead to one larger metacentric or submetacentric chromosome from two smaller acrocentrics. Both centromeres are usually retained in a common centromere region in the fusion chromosome product (b) Human chromosomes (metacentrics, M: submetacentrics, S; and acrocentrics, A) in a metaphase cell.

and in some of the flowering plants. The usual trend is toward reduction in chromosome number as species groups evolve and retention of most of the genome in structurally rearranged chromosomes of the complement. In *Drosophila,* the reduction from a haploid complement of 6 to species with only 3 chromosomes in a set has been examined in some detail from polytene chromosome band comparisons. From these detailed comparisons, it is clear that particular metacentric chromosomes had been derived from centric fusion of particular acrocentrics during evolution since matching band patterns provided unmistakable evidence of these events. The evolutionary significance of aneuploidy by centric fusion lies in the alterations in recombination frequencies and in reduced production of new genotype combinations.

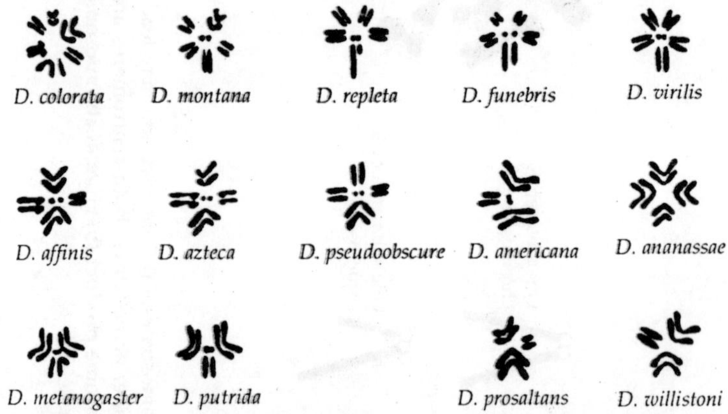

Fig. 16. Chromosome complements of fourteen *Drosophila* species showing the aneuploid series of $n = 6$ to $n = 3$, which has arisen through centric fusions and other events in evolution of the group. The X and Y chromosomes of these male genomes are shown at the bottom of each drawing.

In species with more chromosomes (linkage groups), independent assortment accounts for a high proportion of new genotype combinations in populations. If the same genes are present in fewer chromosomes, crossing over will produce recombinations for these genes. Fewer new genotype combinations will arise by crossing over than by independent assortment. Species with fewer chromosomes tend to be less variable than related species with the same genes in more chromosomes. Some environmental conditions favour low genetic variability. Centric fusions, if they occur, would be retained in such genetically constant populations because of the selective advantages of the slower release of inherent genetic variability.

Index